AF344288

Ferroelectrics UK 2001

Ferroelectrics UK 2001

Edited by

I. M. Reaney and D. C. Sinclair

University of Sheffield

FOR THE INSTITUTE OF MATERIALS

B0764

First published in 2002 by
Maney Publishing
1 Carlton House Terrace
London SW1Y 5DB
for the Institute of Materials

© The Institute of Materials 2002
All rights reserved

ISBN 1-902653-70-X

Typeset in the UK by
Maney Publishing
Printed and bound in the UK at
H. Charlesworth & Co. Ltd.

Contents

Preface

Ferroelectrics are a class of compounds that have existed since the 1930s but it was not until the discovery of a reversible spontaneous polarisation in perovskite structured $BaTiO_3$ during the Second World War that it became of commercial as well as academic interest. Since the discovery of $BaTiO_3$, a large number of perovskite structured materials have been shown to exhibit ferroelectricity. Notably, the solid solution of $PbZrO_3$–$PbTiO_3$ (PZT) has gained world wide acceptance as the material of choice for high performance actuator and sensing applications, including sonar.

In modern life, ferroelectrics are ubiquitous. They are present as the active elements in smart cards, intruder alarms, micromotors and thermal imaging arrays. Ferroelectric capacitors are used in a large number of electronic devices and $BaTiO_3$-based thermally sensitive resistors act as temperature compensating and shut-off devices. These myriad applications drive consistently high quality research in the academic and industrial communities and understanding of ferroelectricity has resulted in an on-going improvement in device performance.

This book is a compilation of the proceedings from Ferroelectrics UK 2001. This annual meeting is designed to bring together industrial and academic researchers and developers in the UK so that they can exchange knowledge and expertise. As a result, the papers contained within deal with a broad range of topics covering the physics, chemistry and materials science/engineering of ferroelectric materials. The meeting attracted delegates from America, Italy and Germany in addition to the UK community. We thoroughly enjoyed hosting the conference in Sheffield and with over 90 delegates in attendance we felt that the meeting was a success and made a worthwhile contribution to the field of ferroelectrics in the UK.

Dr I. M. Reaney and Dr D. C. Sinclair
University of Sheffield

Effect of Epitaxial Strain on the Permittivity of Relaxor Thin Films

G. Catalan, M. H. Corbett, R. M. Bowman and
J. M. Gregg*

*Condensed Matter Physics and Material Science Research Division,
School of Mathematics and Physics, Queen's University of Belfast
Belfast BT7 1NN, United Kingdom*

ABSTRACT

Thin film capacitor structures, containing the lead-based relaxor $Pb(Mg_{1/3}Nb_{2/3})O_3$ (PMN) as the dielectric layer, were made by pulsed-laser deposition on both MgO and $LaAlO_3$ substrates. The dielectric constant of the relaxor was found to be considerably lower than in bulk, while the temperature of the dielectric maximum (T_m) was only slightly depressed, irrespective of the expected in-plane strain induced by the substrate (compressive for $LaAlO_3$ and tensile for MgO). The frequency dependence of T_m was also seen to be more pronounced in thin films than in bulk.

A rationalisation of these differences is presented by consideration of mismatch strain between substrate and film in the context of a semi-empirical modified Landau–Ginzburg–Devonshire model. The predictions of the model are also compared with available experimental results from the literature.

INTRODUCTION

Diffuse phase transitions in relaxors are characterised by a strong frequency dependence of the peak in permittivity as a function of temperature, with the magnitude of the peak decreasing and temperature of the peak (T_m) increasing as frequency increases.[1,2] This complex temperature–frequency behaviour poses a challenge for the description and understanding of relaxor behaviour, particularly in the thin film regime where surfaces, interfaces, size and strain effects further complicate matters.

To date, investigations on relaxor thin films[3–8] have consistently observed a considerable reduction in permittivity compared with bulk/single crystal, in contrast with only relatively little variation in the temperature of the dielectric maximum (T_m) between film and bulk/single crystal. Epitaxial clamping of thin films to rigid substrates is often identified as a major factor in explaining these kinds of differences between film and bulk, and for

---

*Author to whom correspondance should be addressed. Email m.gregg@qub.ac.uk

ferroelectric thin films the influence of epitaxial strain can be evaluated quite successfully in a Landau–Ginzburg–Devonshire (LGD) framework. Although such a treatment has also been attempted to rationalise the effects of strain in relaxor electroceramic films,[7] predicted shifts in T_m differ considerably in magnitude, and even in sign, from those generally observed in experiment.

In this article the authors present dielectric measurements from thin film capacitors with $Pb(Mg_{1/3}Nb_{2/3})O_3$ (PMN) as the relaxor dielectric layer. The authors show that capacitor structures made on both in-plane tensile (MgO) and in-plane compressive $(LaAlO_3)$ (LAO) substrates demonstrate slight negative shifts in T_m, independent of the sign of the in-plane strain. In order to rationalise this, a modified LGD treatment is proposed for the description of the effect of strain on the temperature dependence of the permittivity of relaxors, whose main feature is the assumption of a quadratic, rather than linear, dependence of the first coefficient of the free energy on $(T-T_m)$.

EXPERIMENTAL

Thin film capacitors were made by pulsed laser deposition (PLD) onto single crystal {100} MgO and LAO substrates. $SrRuO_3$ (SRO) was used as the bottom electrode, and gold pads were evaporated through a hard mask to act as top electrodes (contact to two upper gold electrodes gave functional behaviour of two capacitors in series). SRO was deposited at a substrate temperature of 800°C, and a chamber pressure of 0.15 mbar O_2. The substrate temperature was then reduced to 625°C for deposition of PMN. SRO film thickness was ~150–200 nm, and PMN thickness was ~500 nm. The target substrate distance in the system was 75 mm and the energy fluence used was around 1.5 J cm^{-2} at the target surface. A 30 minute post-deposition anneal at 900 mbar O_2 and 600°C was used before removing the specimens from the PLD system.

Electrical characterisation was performed using both HP4263B and HP4284A LCR meters, with specimens in Oxford Instruments cryostats and temperature regulated by Lakeshore 330 Temperature Controllers. Crystallographic information was obtained from Bruker-AXS X-ray diffractometer, and from electron diffraction in a Philips 400T transmission electron microscope.

EXPERIMENTAL RESULTS

θ–2θ X-ray diffraction showed the capacitor structures to be strongly oriented out-of-plane with {100}$_{PMN\ perovskite}$ ‖ {100}$_{SRO\ (pseudocubic\ index)}$ ‖ {100}$_{MgO\,/\,LAO}$ (Fig. 1). A small (~10%) presence of pyrochlore was noted in the capacitors with *best* dielectric properties. This has been seen before for thin film capacitors of PMN-0.3PT made by PLD.[5] Electron diffraction from plan-view transmission electron microscopy specimens showed that the PMN was also strongly oriented in-plane (Fig. 1, inset), implying a relatively coherent interface, and

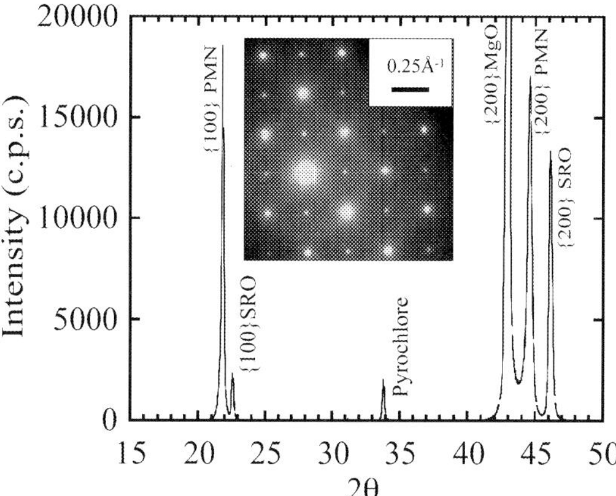

Fig. 1 θ–2θ X-ray diffraction and TEM (inset) from thin film capacitors illustrate strong out-of-plane and in-plane orientation of the thin film, suggesting a degree of coherence/epitaxy with the substrate.

suggesting transfer of epitaxial strain between layers.

Figure 2 shows the dielectric behaviour of PMN grown on both MgO and LAO substrates. Some observations can be made:

(i)   there is a dramatic reduction in the dielectric constant in thin film (bulk single crystals demonstrate dielectric constants between 20,000 and 30,000)

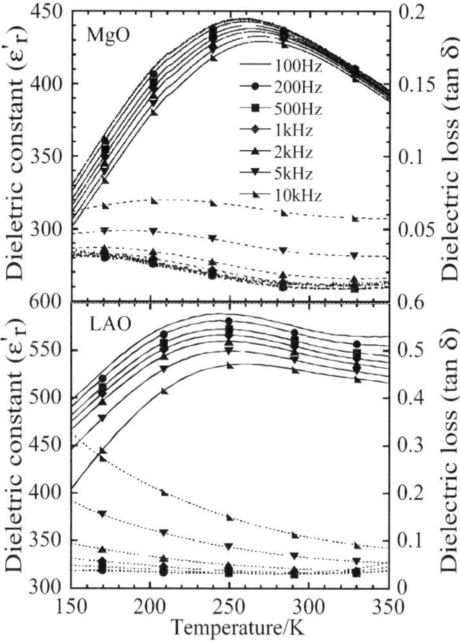

Fig. 2   Relative dielectric constant (ε'_r) and loss (tanδ) as a function of temperature and frequency for PMN thin film capicitors grown on LAO and on Mg.

(ii) T_m for thin film is slightly lower than that of bulk PMN (~260 K for thin film compared with ~268 K for bulk single crystal at 1 kHz)

(iii)   the frequency dependence of T_m is slightly more pronounced in thin films than in bulk single crystals (the difference between T_m at 100 Hz and 10 kHz is ~10 K, whereas in bulk it is ~5 K).

LANDAU–GINZBURG–DEVONSHIRE FORMALISM OF FERROELECTRICS

The bulk pseudocubic lattice parameter for PMN is 4.04 Å, while for LAO it is 3.79 Å, and for MgO 4.2 Å. Therefore, if epitaxial strain is well transferred, the PMN on LAO would be expected to be under in-plane compression, and on MgO under in-plane tension. In ferroelectrics such differences in epitaxial strain would be associated with considerable shifts in the temperature of the dielectric anomaly, with the sign of the shift being positive under compressive in-plane strain and negative for tensile in-plane strain, as shown below.

In the simplest case, Landau–Ginzburg–Devonshire (LGD) formalism[9] allows the thermodynamic potential of a perovskite dielectric thin film to be described as[10,11]

$$\Delta G = \left(\frac{1}{2}\alpha - Q_{i3}X_i \right)P_3^2 + \frac{1}{4}\beta P_3^2 - \frac{1}{2}s_{ij}\left(X_i X_j \right) \qquad (1)$$

where a and b correspond to the linear and non-linear terms of the inverse permittivity, and s_{ij}, X_i and Q_{i3} are the elastic compliances, the stress tensor and the electrostriction tensor in Voigt notation; P_3 is the out-of-plane polarisation. The inverse permittivity is the second derivative of the free energy with respect to the polarisation

$$\chi_3' = \frac{\delta^2 \Delta G}{\delta P_3^2} = \alpha - 4Q_{13}\frac{Y}{1-v}x_m + 3\beta P_3^2 \qquad (2)$$

where the two components of the in-plane elastic stress (x and y directions) have been related to the mismatch strain, x_m, assuming epitaxial 'cube on cube' growth, and Y and v are the Young's modulus and Poisson's ratio of the film, respectively. For conventional ferroelectrics, α is usually expanded in a power series of $(T–T_C)$

$$\alpha = \frac{1}{C\varepsilon_0}(T - T_C) + O(T - T_C)^2 \qquad (3)$$

which leads directly to Curie–Weiss behaviour, where C is the Curie constant. The linear dependence of α on $(T–T_C)$ also introduces the change of sign necessary to stabilise the ferroelectric state below T_C. Substituting equation (3) into equation (2) shows that strain shifts the critical temperature of a ferroelectric by

$$\Delta T_C = 4C\varepsilon_0 Q_{13}\frac{Y}{1-v}x_m \qquad (4)$$

Since Q_{13} is always negative, it can readily be seen that the shift in the critical temperature depends on the sign of the in-plane strain: T_C will decrease for the tensile case and will increase for the compressive case. In contrast, most observations for relaxor thin films show T_m to decrease slightly under both tensile and compressive strain.[4,5] It may be that the shift in T_m for relaxor thin films is not caused by strain, or else that a different formalism is needed to describe the effect of biaxial strain on relaxors. In the next section this possibility is explored.

MODIFIED LANDAU–GINZBURG–DEVONSHIRE TREATMENT FOR RELAXORS

It has been proposed[7] that equation (4) can be used for relaxors if T_C is substituted by T_m. This approach relies on the assumption that the temperature dependence of the inverse permittivity of relaxors can be described by a linear Curie–Weiss law, as in equation (3). However, Curie–Weiss behaviour is known to break down for relaxors below the temperature at which $P_{rms} \neq 0$ (T_{Burns}) (Ref. 12) and thus equation (3) does not accurately describe the shape of the dielectric peak. Furthermore, if a linear temperature dependence of α on $(T-T_m)$ was implemented in the LGD free energy expression, the change of sign of α below T_m would imply a transition to a stable ferroelectric state at T_m, which is wrong.

In ferroelectrics with diffuse phase transitions, the permittivity is instead found to be well described by $\dfrac{1}{\varepsilon} - \dfrac{1}{\varepsilon_m} = C^{-1}(T - T_m)^{\gamma}$, with $\gamma = 2$ for lead-based relaxors,[13–18] being usually attributed to a Gaussian distribution of critical temperatures and/or sizes of the polar nano-regions.[19] In fact, it is usual to relate the quadratic temperature dependence of the dielectric constant to the first term of the Taylor expansion of a Gaussian curve

$$\frac{1}{\varepsilon} - \frac{1}{\varepsilon_m} = \frac{1}{2\varepsilon_m \delta^2}(T - T_m)^2 \tag{5}$$

To illustrate this quadratic dependence, the dielectric constant has been extracted along the [001] direction for single-crystal PMN from the literature,[20] presented in Fig. 3 for 100 kHz. The proportionality to $(T-T_m)^2$ also seems to be valid for temperatures below T_m (Fig. 3), consistent with the predictions of the Edwards–Anderson classic model of glasses.[21] The data was well fitted by straight lines both above and below T_m, with $\delta \sim 36$ K above, and $\delta \sim 25$ K below T_m (Ref. 22).

Empirically, then, the inverse permittivity at a specific frequency, is well described by the quadratic fit presented in equation (5). In order to also describe the frequency dependence of the permittivity, the frequency dependent parameters $T_m(f)$ and $\varepsilon_m(f)$ must also be determined. At intermediate frequencies,[23] $T_m(f)$ is usually described using the Vögel–Fulcher expression $\left(\tau = \tau_0 \exp\left[U/_{T_m - T_f}\right]\right)$ (Ref. 24); $\varepsilon_m(f)$, on the other hand, was found to be well described by a quadratic relationship analogous to equation (5) (inset of Fig. 3)

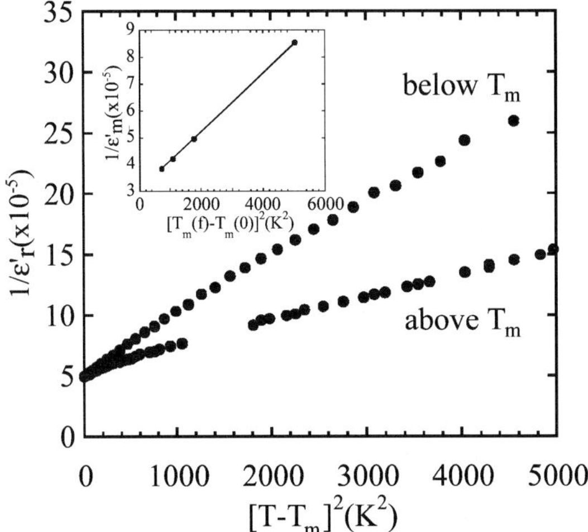

Fig. 3   The linear dependence of the reciprocal dielectric constant of a PMN single crystal (100 kHz) on $(T–T_m)^2$ above and below T_m. Inset: inverse of the permittivity maxima (circles) and best fit (solid line) to equation (6).

$$\frac{1}{\varepsilon_m(f)} - \frac{1}{\varepsilon_m(0)} = \frac{1}{2\varepsilon_m(0)\delta^2}\left[T_m(f) - T_m(0)\right]^2 \tag{6}$$

where (f) and (0) refer to values at finite and zero frequency, respectively. The zero-frequency values ($\varepsilon_m(0)$=32800, $T_m(0)$ = 235 K, δ = 36 K) were extracted as fitting parameters to the single crystal data plotted in the inset of Fig. 3.

The parameters τ_0, U, $T_m(0)$, $\varepsilon_m(0)$ and C (or δ) fully characterise $\varepsilon\,(T,f)$ for single-crystal PMN along [001]. Since the inverse permittivity is the second derivative of the free energy with respect to the polarisation in equation (2), the dielectric constant must derive from the coefficients in the LGD thermodynamic potential. For P=0, and in the absence of strain, equation (2) indicates that the low-field inverse permittivity corresponds to the first coefficient of the expansion, i.e, $\chi(T,f)=\alpha(T,f)$. Accordingly, α must have a quadratic, rather than linear, dependence on $(T–T_m)$. Once $\alpha(f,T)$ has been extracted from the unperturbed inverse permittivity of a single crystal, equation (2) allows the expected shape of the dielectric peak of PMN under mismatch strain to be described.

The absolute value of the dielectric constant calculated from equation (2) is proportional to x_m, and is thus very sensitive to the influence of the substrate. In contrast, shifts in T_m were found to be essentially independent from the absolute value, or even the sign, of the strain. The temperature of the dielectric maximum under strain is calculated differentiating the inverse permittivity with respect to temperature at $T_m{'}$

$$\left.\frac{\delta\chi'}{\delta T}\right|_{T_m'} = 0 = \frac{1}{\varepsilon_0\varepsilon_m}\frac{\left[T_m' - T_m\right]}{\delta^2} - 4Q_{13}\frac{Y}{1-\nu}\frac{\delta x_m}{\delta T} \tag{7}$$

assuming that Q_{13}, Y and ν do not change substantially with temperature in the region of the permittivity peak, as observed[2] and only the mismatch strain changes, due to differential thermal expansion

$$\frac{\delta x_m}{\delta T} = \frac{\delta}{\delta T}\left(\frac{a\| - a_{PMN}}{a_{PMN}}\right) \tag{8}$$

where $a\|$ is the in-plane lattice parameter of the PMN film, and a_{PMN} is the unperturbed lattice parameter of PMN, which is relatively constant below ~550 K (Ref. 25). Due to clamping, $a\|$ is constrained to follow the same thermal evolution as the single-crystal substrate

$$a\| = a\|^0\left[1 + \lambda_{subs}\left(T - T_0\right)\right] \tag{9}$$

where λ_{subs} is the coefficient of thermal expansion of the substrate, and $a\|^0$ is the film in-plane lattice parameter at the deposition temperature T_0.

Thus, the shift in T_m is

$$\Delta T_m = 4\delta^2\varepsilon_m\varepsilon_0 Q_{13}\frac{Y}{1-\nu}\frac{a\|^0}{a_{PMN}}\lambda_{subs} \tag{10}$$

COMPARISON TO EXPERIMENTAL DATA AND DISCUSSION

The modelled permittivity of a relaxor single crystal, using the approach described in the previous section, is shown in Fig. 4. In the same figure, the predicted shape of the relative dielectric permittivity for a relaxor film under tensile epitaxial strain $x_m = 0.04$, corresponding to the lattice mismatch between PMN and MgO, is shown. The qualitative and quantitative similarity with our experimental data (Fig. 1) is notable.

The authors do not believe, however, that the full extent of the decrease in dielectric constant with respect to bulk is due purely to strain: if it was so, then films grown under compressive in-plane strain should display increased dielectric constant, which is not the case. Also, there is likely to be some strain relaxation across the film, and therefore the mismatch strain used in the calculations is a maximal.

The shifts in T_m predicted by equation (10) for PMN and PMN–PT films grown on different substrates are shown on Table 1. As noted in the previous section, there are two different values for the 'diffusivity' coefficient δ, depending on whether the low temperature or the high temperature fit is used. In the table, the predicted shifts for both extremes are specified. The value of δ for PMN–0.1PT single crystals could not be found in the literature, and the same value has been used as for pure PMN (25 K below T_m, 36 K above T_m); for PMN-0.3 PT, $\delta=24$ K was found.[27] In all calculations $Y=100$ GPa (Ref. 2) and $\nu = 0.3$ (Ref. 2) has been assumed as for pure PMN. The experimentally measured shifts in T_m for PMN-PT thin films have been included in the table for comparison, showing reasonable quantitative agreement with these calculations.

Qualitative analysis of the shifts predicted by equation (10) also reveals several interesting features. Firstly, since compositions close to the morphotropic phase boundary have larger permittivities, greater shifts in T_m should be expected. This would be consistent with the observation that, as the composition of the relaxor films gets closer to the morphotropic phase boundary (and thus the intrinsic bulk permittivity is higher), so the experimentally measured shift in T_m increases from ~–5 K for pure PMN, to ~–10 K for PMN–0.1 PT, to ~–35 K for PMN–0.3PT.

Also, as the predicted shift in T_m is proportional to ε_m, the low-frequency, higher valued dielectric peaks, will shift more than the lower, high-frequency ones: consequently the frequency dependence of T_m will be more pronounced in thin films than in bulk, an experimental observation that was noted above.

Importantly, equation (10) predicts that T_m can only decrease, regardless of the sign of the in-plane strain. This prediction is in sharp contrast with conventional ferroelectric phenomenology, which predicts an increase in T_C when the in-plane strain is compressive. Both these results, and the results by Maria et al.[5] using LAO substrates (in-plane compressive strain) show decreased T_m with respect to bulk values, in agreement with this model.

CONCLUSIONS

An empirical model based on LGD formalism has been used to rationalise the effect of biaxial strain on the permittivity of relaxor PMN. The distinctive feature of the treatment

Table 1   Comparison between modelled and experimentally measured shifts in T_m for PMN–PT films grown on different substrates. The two calculated shifts for each composition correspond the values of δ below and above T_m respectively. Brackets indicate the references from which the data has been extracted.

Film / Substrate	Bulk ε_m	Q_{13} (m^{4}C^{-2})	λ_{subs} (K^{-1})	$\Delta T_m^{\text{model}}$	$\Delta T_m^{\text{experiment}}$
PMN / MgO	22000[20]	-0.96[16]	1.2 x 10^{-5}	-9 K/-18 K	-5 K
PMN/LAO	22000[20]	-0.96[16]	1.0 x 10^{-5}	-8 K/-16 K	-20 K
PMN / Si	22000[20]	-0.96[16]	6.7 x 10^{-6}	-4 K/-8 K	-8 K[3]
PMN-0.1PT / MgO	27000[14]	-1.0[26]	1.2 x 10^{-5}	-11 K/-22 K	-10 K[7]
PMN-0.1PT/LAO	27000[14]	-1.0[26]	1.0 x 10^{-5}	-10K/-20K	+110 K*[7]
PMN-0.3PT/LAO	34500[28]	-3.5[28]	1.0 x 10^{-5}	-38 K	-35 K[5]

* A positive shift in T_m is in contradiction to the predictions of the authors' model. The experimental results of Nagarajan et al.,[7] however, also contradict the experimental trends generally observed in other works, and should be taken with some caution, as the temperature and frequency dependence of the dielectric constant, essential to establish relaxor behaviour, was not shown; dielectric loss was not reported either, which means that space-charge effects cannot be ruled out.

was the assumption of a quadratic, rather than linear, temperature dependence of the first coefficient, α, in the thermodynamic potential, consistent with the empirical observation that the small-field permittivity has a quadratic dependence on $T\text{-}T_m$.

The model was found to provide an accurate fit to the permittivity of relaxors, with and without strain, as a function of temperature and frequency, and was particularly successful in reproducing the behaviour of T_m in thin films, accounting for

(i)   a small decrease in T_m regardless of the absolute value or the sign of the in-plane strain

(ii)   more pronounced frequency dependence of T_m in thin films than in bulk and

(iii)   bigger shifts for films with compositions closer to the morphotropic phase boundary, due to the proportionality between the predicted shift and ε_m.

ACKNOWLEDGEMENTS

This work was funded by the Engineering and Physical Sciences Research Council.

REFERENCES

1.   G. A. Smolenskii, V. A. Isupov, A. I. Agranoskaya and S. N. Popov, *Sov. Phys. Solid State*, 1961, 2584.
2.   Landolt-Börnstein, *Numerical Data and Functional Relationships in Science and Technology*, Vol. 16, K.-H. Helwege and A.M. Hellwege, eds, Springer-Verlag, 1981.
3.   C. Tantigate, J. Lee and A. Safari, *Appl. Phys. Lett.*, 1995, **66**, 1611.
4.   Z. Kighelman, D. Damjanovic, A. Seifert, L. Sagalowicz and N. Setter, *Appl. Phys,. Lett.* 1998, **73**, 2281; Z. Kighelman, D. Damjanovic and N. Setter, *J. Appl. Phys.*, 2001, **89**, 1393.
5.   J. P. Maria, W. Hackenberger and S. Trolier-McKinstry, *J. Appl. Phys.*, 1998, **84**, 5147.
6.   G. Catalan, M. H. Corbett, R. M. Bowman and J. M. Gregg, *Appl. Phys. Lett.*, 1999, **74**, 3035.
7.   V. Nagarajan et al., *Appl. Phys. Lett.*, 2000, **77**, 438.
8.   G. R. Bai et al., *Appl. Phys. Lett.*, 2000, **76**, 3106.
9.   F. Devonshire, *Philos. Mag.*, 1949, **40**, 1040; *Philos. Mag.* 1951, **42**, 1065.
10.   G. A. Rossetti Jnr, L. E. Cross and K. Kushida, *Appl. Phys. Lett.*, 1991, **59**, 2504.
11.   N. A. Perstev, A. G. Zembilgotov and A. K. Tagantsev, *Phys. Rev. Lett.*, 1998 **80**, 1988.
12.   D. Viehland, S. J. Jang, L. E. Cross and M. Wuttig: *Phys. Rev. B*, 1992, **46**, 8003.
13.   G. A. Smolensky, *J. Phys. Soc. Jpn*, 1970, **28** (suppl.), 26; G. A. Smolenskii, *Ferroelectrics and Related Materials*, Gordon and Breach, 1984.
14.   T. R. Shrout and J. Fielding, *Proc. 1990 IEEE Ultrasonics Symp.*, the Institute of Electrical and Electronics Engineers, 1990, 711.
15.   S. Chattopadhay, P. Ayyub, V. R. Palkar and M. Multani, *Phys. Rev. B*, 1995, **52**, 13177.
16.   K. Uchino, S. Nomura, L. E. Cross, S. J. Jang and R. E. Newnham, *J. Appl. Phys.*, 1980, **51**, 1142.
17.   G. H. Blackwood and M. A. Ealey, *Smart Mater. Struct.* **2**, 1993, 124.
18.   A. K. Tagantsev, *Phys. Rev. Lett.*, 1994, **72**, 1100.
19.   A. J. Bell, *J. Phys. Condens. Matter.*, 1993, **5**, 8773.

20. A. Levstik, Z. Kutnjak, C. Filipic and R. Pirc, *Phys. Rev. B,* 1998, **57**, 11204.

21. S. F. Edwards and P. W. Anderson, *J. Phys. F,* 1975, **5**, 965.

22. Below T_m there is a further, very slight, change of slope around T≈235–240 K, the freezing temperature reported for this set of data (Ref. 20).

23. E. V. Colla, E. Yu Koroleva, N. M. Okuneva and S. B. Vakhrushev, *J. Phys.: Condens. Matter,* 1992, **4**, 3671.

24. D. Viehland, S. J. Jang, L. E. Cross and M. Wuttig, *J. Appl. Phys.,* 1990, **68**, 2916.

25. J. Zhao, A. E. Glazounov, Q. M. Zhang and B. Toby, *Appl. Phys. Lett.,* 1998, **72**, 1048.

26. Q. M. Zhang and J. Zhao, *Appl. Phys. Lett.* 1997, **71**, 1649.

27. T. R. Shrout, Z. P. Chang, N. Kim and S. Markgraf, *Ferroelectrics Lett.,* 1990, **12**, 63.

28. S.-E. Park, private communication.

Comparison of Ferroelectric, Piezoelectric and Relaxor Materials via Simulation of Ultrasonic Transducers

A. Cochran and K. J. Kirk

Microscale Sensors, University of Paisley, Division of Electronic Engineering and Physics, Paisley PA1 2BE, UK

ABSTRACT

Most ultrasonic transducers are based on polycrystalline piezoelectric ceramics, such as $Pb(Ti_{1-x}Zr_x)O_3$ (PZT), but some ferroelectric materials, such as single crystal $LiNbO_3$ (LNO), are also still used. Relatively recently, these classes of material have been joined by relaxor-based single crystal piezoelectric materials, such as $Pb(Zn_{1/3}Nb_{2/3})O_3$–$PbTiO_3$ (PZN–PT), now available commercially in small quantities. Experimental comparison of materials in all three classes is time consuming and expensive. Instead, here, we present comparative material parameters, including voltage and strain constants, permittivity and electromechanical coupling, then, using results from well-established simulation techniques, we illustrate how these affect the performance of ultrasonic transducers in different configurations. We also discuss practical limitations. We conclude that no one class of material outperforms the others in all applications and that care must be taken in material selection by considering a full range of parameters which must also be incorporated in the transducer design process. We also comment on desirable practical properties for future materials for a range of applications of ultrasound.

INTRODUCTION

Ultrasonic transducers have five main markets: non-destructive testing (NDT), biomedical imaging and therapy, underwater sonar, industrial processing and remote sensing. The materials used for electromechanical transduction are usually small but critical components in ultrasonic systems in all these markets.

There are now three distinct generations of bulk electromechanically-active materials. Ferroelectric single crystals, including quartz and $BaTiO_3$, were the first materials to be used in ultrasonic transducers.[1] In this paper, $LiNbO_3$ (LNO) is used as an example. Piezoelectric ceramics were developed after ferroelectric single crystals. They come in many variations, mostly Pb-based; the example used here is $Pb(Zr,Ti)O_3$, specifically PZT–5H.[2] Finally, electrostrictive relaxors have recently been developed.[3] These are single crystal compounds, such as $Pb(Mg_{1/3}Nb_{2/3})O_3$ (PMN) and $Pb(Zn_{1/3}Nb_{2/3})O_3$–8% $PbTiO_3$ (PZN–PT), which is considered here.

The effects of these materials on transducer performance can be illustrated by comparing simulated characteristics of a typical NDT transducer and single elements of a biomedical imaging array and a sonar array made from each of the three materials. In the next section we outline transducer design. Then we compare some of the key parameters of the materials of relevance to transducers. We next briefly outline our simulation approach, then present results from it. Finally we draw conclusions regarding the suitability of the three materials for use in ultrasonic transducers.

TRANSDUCER DESIGN

Each transducer is representative of its type. All three are assumed to operate in a quasi-resonant mode in which the thickness of the active element is half the ultrasonic wavelength at the operating frequency. Although not every detail of the transducers is presented here, "best practice" has been used throughout their design.

NDT Transducer[4]

This transducer has a single element to transmit and receive ultrasound at a typical NDT frequency of 5 MHz, as shown in Fig. 1. The key geometrical issue is that the width-to-thickness ratio of the active material is much more than one. Hence, plane waves propagate through its thickness and conventional bulk material parameters can be used to simulate its behaviour. The transducer is operated on an aluminium specimen. Since bandwidth is important, a "backing block" with a moderately high acoustic impedance is included for mechanical damping.

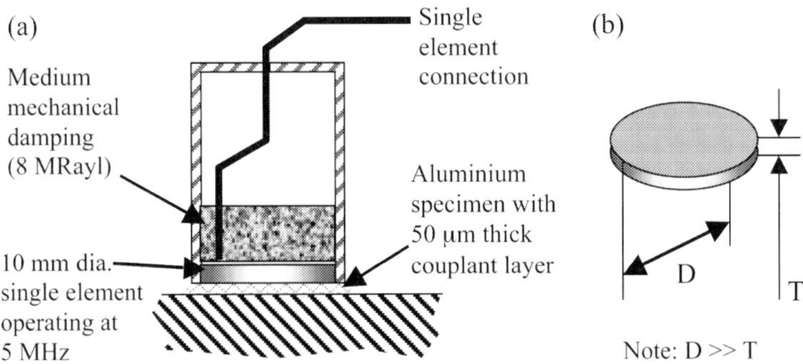

Fig. 1   (a) schematic diagram of a single element transducer for NDT and (b) the shape of the element.

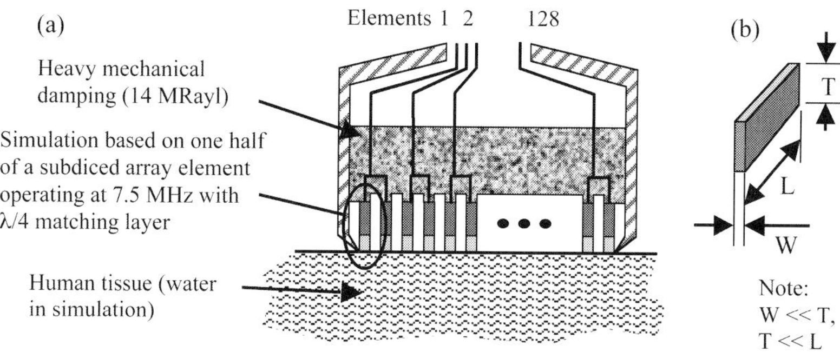

Fig. 2   (a) schematic diagram of a biomedical array and (b) the shape of one "sub-diced" section of element.

Biomedical Array[5]

A biomedical imaging array comprises a set of adjacent elements, here at a typical operating frequency of 7.5 MHz. For high quality images, the centre-to-centre spacing of the elements must be less than half the propagation wavelength in tissue. This gives a width-to-thickness ratio undesirably close to one when the gap between the elements is included, so each element is assumed to be sub-diced into two narrower "planks" 25 mm wide, as shown in Fig. 2. These can be modelled with equivalent material parameters derived from bulk figures when geometry is taken into account.[6] In the simulations, water has been used to mimic human tissue and a backing block with a high acoustic impedance is included because operating bandwidth is critical in imaging. A single quarter-wavelength thick "matching layer" is also used to improve performance.

Sonar Array[7]

The type of sonar array considered here is similar to the biomedical one, but operating at a frequency of 200 kHz, as illustrated in Fig. 3. High performance sonar transducers are made

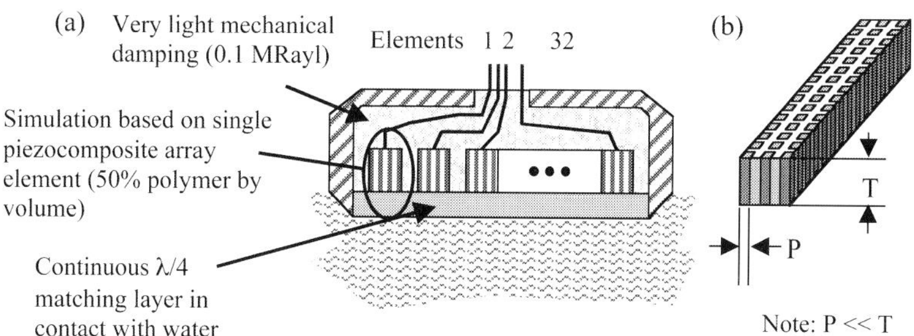

Fig. 3   (a) schematic of a sonar array and (b) the shape and composition of each element (dark grey – active pillars; light grey – passive matrix).

with piezocomposite material,[8] typically comprising tall, narrow pillars of an electromechanically active material supported by a passive polymer matrix. Again, equivalent material parameters can be used to model the behaviour of each array element.[9] A quarter-wavelength thick matching layer improves performance but since mechanical damping is difficult at low frequencies, the elements are assumed to be supported by a material with a low acoustic impedance.

MATERIALS

Selected parameters of the three example materials are shown in Table 1. Some of the figures are approximate because data in the literature differ and manufacturing tolerances can be as high as ±20%. In any case, the discussion here, and choices made in transducer design, will typically be based on differences larger than that.

First note that d_{33} for LNO is two orders of magnitude less than for PZT–5H, itself approximately a quarter that of PZN–PT. There are much smaller differences in g_{33}. The relative permittivities of LNO under constant stress and strain are much smaller than for the other materials. PZT–5H has a higher permittivity than PZN–PT at constant stress and PZN–PT is higher at constant strain, differences which could affect performance under different operating conditions.

Electromechanical coupling coefficients make it easy to compare materials[10] independently of parameters, such as permittivity, which vary widely. They describe the conversion of energy stored in the material into electrical or mechanical work and are very useful for the

Table 1   Key materials parameters for comparison for ultrasonic transducer design.

Parameter	Symbol	Units	LNO	PZT–5H	PZN–PT
Piezoelectric strain constant	d_{33}	pm/V	5.88	594	2200
Piezoelectric voltage constant	g_{33}	mV/N	22.3	19.5	48.7
Relative permittivity at constant stress	ε_{33}^{S}		29.0	1470	560
(and strain)	(ε_{33}^{T})		(29.8)	(3400)	(5100)
Electromechanical coupling coefficient	k_{31}		-0.024	-0.386	-0.593
	k_{33}		0.162	0.747	0.940
	k_{T}		0.162	0.512	0.483
	k_{P}		-0.037	0.648	-0.927
Mechanical Stiffness	c_{33}^{E}	MN.cm^{-2}	24.5	11.7	10.5
Mechanical quality factor	Q		>1000	65	40
Density	ρ	kg.m^{-3}	4640	7500	8000
Thickness mode velocity	v	m.s^{-1}	7360	4600	4140
Specific acoustic impedance	Z	MRayl	34.2	34.5	33.1
Curie Temperature	T_c	°C	1210	195	150

ultrasonic transducer designer. The four most common coefficients are shown in Table 1. In general, the figures for LNO are smaller than for the other two materials, by factors as large as 25. k_{33} corresponds to length-wise resonance of a thin solid cylinder with electrodes on its ends, poled along its axis. It is larger for PZN–PT than PZT–5H. k_T corresponds to the fundamental thickness resonance of a plate with electrodes on its major surfaces, poled through its thickness, and is similar for the two materials. The importance of these comparisons is illustrated later.

The other parameters in Table 1 cover a range of properties. LNO is stiffest and has the highest quality factor, but neither these nor its lower density is critical in transducer design. However, its very high thickness mode velocity means that it is thicker for a given operating frequency, potentially useful for high frequencies. LNO's other key parameter is its very high Curie temperature (T_c). PZT–5H and PZN–PT are broadly similar across the same range of parameters, despite the PZT–5H being a ceramic and PZN–PT a single crystal, a difference which may itself affect transducer manufacture.

SIMULATION

The simulated results are based on solutions of the one-dimensional wave equation in the transducers. As outlined above, the geometry of the NDT transducer matches one-dimensional, plane wave propagation closely. The geometries of the elements in the biomedical and sonar arrays can also be dealt with using one-dimensional solutions in which the shapes of the elements are taken into account via equivalent material parameters calculated from the bulk parameters and the particular geometry.

Equivalent experimental results are not presented here. This is because simulation has a very well-established history in ultrasonic transducer design[11] and the computer code used to obtain the present results has been extensively validated with experimental results from the literature. Therefore, rather than using modelling to enhance fundamental understanding, it is used here to explore the effects of different materials on transducer behaviour, unconstrained by experimental limitations, such as laboratory time, manufacturing expense and device repeatability.

RESULTS

Results are presented for each transducer in turn, covering four characteristics. The first is the electrical impedance of the element under operating conditions. Typically, a transducer is operated on transmission at the frequency of a local minimum in its electrical impedance spectrum, and on reception at the frequency of the corresponding local impedance maximum. A low minimum impedance is most effective for transmission with low output-impedance circuitry, and a high maximum works well with a high input-impedance receiving voltage amplifier. For the simulations here, the output-impedance of the transmitting circuit was

Table 2   Key performance parameters for the NDT transducer.

		LNO	PZT–5H	PZN–PT
Electrical impedance magnitude				
Local Minimum (at frequency)	Ω (MHz)	1130 (4.94)	7.75 (4.56)	19.9 (4.64)
Local Maximum (at frequency)	Ω (MHz)	1150 (5.25)	20.3 (5.62)	44.8 (5.59)
Transmission characteristics				
Relative amplitude	dB	-28.7	0.00	-4.00
Quality factor, Q_{TX}		5.90	4.05	4.23
Reception characteristics				
Relative amplitude	dB	0.00	-10.5	-7.35
Quality factor, Q_{RX}		5.54	5.17	4.88
Transmit–receive characteristics				
Relative amplitude	dB	-16	0.00	-0.327
Quality factor, Q_{RX}		9.00	4.86	5.28

assumed to be 0.1 W and the input impedance of the receiver was 1 mW. The separation between the transmission and reception frequencies is also important for practical operation.

The other three key characteristics are transmission behaviour, reception behaviour and transmit–receive behaviour. Each is described here in terms of peak signal magnitude in the frequency domain, which should be as high as possible, and quality factor (Q), which is indicative of operating bandwidth and should be as low as possible for the majority of applications. Impulse drive signals are assumed, voltage being applied to the transducer via the electrical circuits for transmission and transmit–receive operation and pressure on the front face for reception.

NDT Transducer

Data for the NDT transducer are given in Table 2. The transducer made with LNO has a much higher electrical impedance than the other two, suggesting that it will have poorer transmission characteristics but better reception ones.

As expected, the pressure output from LNO is much less than for PZT–5H, indicating poorer transmission performance. However, LNO generates the highest voltage across the 1 MW receiver input resistance on reception, attributed to its higher impedance (matching the receiver resistance better) and the relatively small discrepancy in g_{33}. The data also show that the pressure output from PZT–5H is 4 dB higher than from PZN–PT, but that Q_{TX} is also slightly higher, an undesirable characteristic. The responses are otherwise very similar. On reception, PZN–PT produces a signal approximately 3 dB better than PZT–5H and Q_{RX} is slightly better.

The transmission and reception behaviour together define the transmit–receive results so it is not unexpected that the signal magnitudes for PZT–5H and PZN–PT are very similar. In fact, both the signal magnitude and $Q_{TX/RX}$ for PZT–5H are better than for PZN–PT. This can be explained by the fact that $Q_{TX/RX}$ is a function of Q_{TX} and Q_{RX}, but also of the difference in frequencies of peak output for transmission and reception, largely a result of the material parameters. Although each transducer has been designed to maximise transmit–receive output at a frequency close to the specification, the transmission and reception frequencies can be quite different, as indicated by the frequencies of impedance minimum and maximum, leading to unexpected values for $Q_{TX/RX}$. The transmit–receive performance of the LNO is significantly poorer than the other two materials: signal magnitude is approximately 16 dB lower and $Q_{TX/RX}$ is much higher, principally because of the small difference between transmission and reception frequencies, indicating a smaller operating bandwidth.

Biomedical Array

The data for the biomedical array are given in Table 3. Here, the very much smaller elements have much higher electrical impedances than for NDT. In addition, the increased mechanical damping reduces the difference between the local impedance minimum and maximum for each material. In fact, local features do not exist at all for LNO, because of the heavy damping and weak electromechanical coupling and the element impedance at the operating frequency is approximately 1 MW, very high for transmission. The relevant impedances of PZT–5H and PZN–PT are much lower, in the region of 1–10 kW, relatively convenient figures for practical operation.

On transmission, the performance of the LNO is poorest in terms of both pressure amplitude and Q_{TX}. PZT–5H produces a marginally high pressure, but Q_{TX} is significantly higher. On reception, LNO gives the highest voltage, but Q_{RX} is by far the worst. The characteristics of PZT–5H are appreciably worse than for PZN–PT. Finally, the transmit–receive signal

Table 3   Key performance parameters for the biomedical array element.

		LNO	PZT–5H	PZN–PT
Electrical impedance magnitude				
Local Minimum (at frequency)	kΩ (MHz)	n/a	2.05 (5.27)	2.12 (2.86)
Local Maximum (at frequency)	kΩ (MHz)	n/a	3.47 (9.38)	6.19 (10.8)
Transmission characteristics				
Relative amplitude	dB	-32.6	0.00	-0.48
Quality factor, Q_{TX}		2.58	1.66	0.72
Reception characteristics				
Relative amplitude	dB	0.00	-8.34	-5.49
Quality factor, Q_{RX}		2.28	1.80	1.30
Transmit–receive characteristics				
Relative amplitude	dB	-26.2	-2.15	0.00
Quality factor, $Q_{TX/RX}$		3.44	2.33	1.68

Table 4   Key performance parameters for the sonar array element.

		LNO	PZT–5H	PZN–PT
Electrical impedance magnitude				
Local Minimum (at frequency)	kΩ (kHz)	n/a	1.30 (144)	1.12 (112)
Local Maximum (at frequency)	kΩ (kHz)	n/a	6.62 (256)	16.5 (286)
Transmission				
Relative amplitude	dB	-34.2	-0.87	0.00
Quality factor, Q_{TX}		3.42	1.84	2.52
Reception				
Relative amplitude	dB	0.00	-6.60	-2.51
Quality factor, Q_{RX}		2.69	4.22	6.27
Transmit–receive				
Relative amplitude	dB	-22.1	-0.429	0
Quality factor, $Q_{TX/RX}$		4.82	1.75	1.08

amplitude and $Q_{TX/RX}$ for PZN–PT clearly surpass the figures for PZT–5H, with LNO much worse.

Sonar Array

The results for the sonar array element are shown in Table 4. In this case, the active materials have been used as part of the piezocomposite element. Key material parameters in this case are k_{31} and k_{33}. These are much lower for LNO than PZT–5H and PZN–PT, explaining the absence of a local impedance minimum and maximum. The figures for the other two materials are within a reasonable range for instrumentation.

On transmission, PZN–PT produces the highest pressure output, though by less than 1 dB, whilst Q_{TX} is best for PZT–5H. As usual, LNO lags far behind in terms of pressure output. However, it is still best on reception, though by a relatively small margin. When it comes to transmit–receive operation, PZN–PT has the best combination of characteristics, with a slightly better signal amplitude than PZT–5H, but $Q_{TX/RX}$ is almost twice as good. The reason is that PZN–PT has a very high value of k_{33}. This makes it particularly suitable for use as the pillars of active material in a piezocomposite. It also leads to a large separation in frequency between transmission and reception, which translates into a very low value of $Q_{TX/RX}$ when both are taken into account. Whilst the pressure and voltage amplitudes are relatively similar to PZT–5H in the example given here, careful design could trade off quality factor for signal amplitude, resulting in much higher overall sensitivity than PZT–5H for the same operating bandwidth.

CONCLUSIONS

Using well-known simulation techniques, we have carefully compared the performance of three materials for electromechanical transduction when used in three different types of ultrasonic transducer. Each material can be considered characteristic of its class and, although the results of simulation will not correspond exactly with reality, it is the differences between them that are important here.

In a single-element NDT transducer made with a single-phase active material rather than a composite, there is little to choose between the piezoelectric ceramic PZT–5H and the relaxor PZN–PT in terms of performance. However, the ceramic is presently orders of magnitude less expensive, and is therefore the material of choice. The performance of LNO is much poorer, but its high T_c makes it suitable for application at elevated temperatures[12] at which the other two materials would not survive.

The results for the biomedical array element suggest that a small increase in transmit–receive signal output could be gained by using PZN–PT rather than PZT–5H. More importantly, $Q_{TX/RX}$ is almost 30% lower for PZN–PT than PZT–5H, indicating a significant potential improvement in operating bandwidth. This may be a key issue in new biomedical systems utilising second harmonic imaging.[13] In biomedicine, the high T_c of LNO is unlikely to be important, and it will not be a material of choice.

LNO can again be discounted as the active material in the composite sonar array element that has been simulated. In this case, the transmit–receive signal amplitudes of PZT–5H and PZN–PT are almost identical. However, $Q_{TX/RX}$ is almost 40% more for PZN–PT than for PZT–5H, again indicating that it merits consideration if bandwidth is critical, since its higher cost may be a very small fraction of the cost of a sonar system.

REFERENCES

1. D. A. Berlincourt, D. R. Curran and H. Jaffe, 'Piezoelectric and piezomagnetic materials and their function in transducers', in *Physical Acoustics, Vol. 1A*, W. P. Mason (ed.), Academic Press, 1964.
2. Morgan Electro Ceramics Ltd, *Piezoelectric Ceramics Data Book*, Southampton, UK, 2001.
3. S.-E. Park and T. R. Shrout, 'Characteristics of relaxor-based piezoelectric single crystals for ultrasonic transducers', *IEEE Trans. Ultrason. Ferroelec. Freq. Cont.*, 1997, **44**(5), 1140–1147.
4. J. Krautkramer and H. Krautkramer, *Ultrasonic Testing of Materials*, 4th edn, Springer Verlag 1990, 187–204.
5. W. R. Hedrick, D. L. Hykes, and D. E. Starchman, *Ultrasound Physics and Instrumentation*, Mosby, 1994, 96–111.
6. G. Hayward and D. Gillies, 'Block diagram modeling of tall, thin parallelipiped piezoelectric structures', *J. Acoust. Soc. Am.*, 1989, **86**(5), 1643–1653.
7. D. Stansfield, *Underwater Electroacoustic Transducers*, Bath University Press, 1991, 18–24.
8. A. Shaulov, W. A. Smith and B. Singer, 'Composite piezoelectrics for ultrasonic transducers', *IEEE Trans. Ultrason. Ferroelec. Freq. Contr.*, 1986, **33**(6), 812–821.

9.   W. A. Smith and B. A. Auld, 'Modeling 1-3 composite piezoelectrics: thickness-mode oscillations', *IEEE Trans. Ultrason. Ferroelect. Freq. Contr.*, 1991 **48**, 40–47.

10. IEEE, 'IEEE Standard on Piezoelectricity', ANSI/IEEE Standard 1976–1978, 1979.

11. G. K. Lewis, 'A matrix technique for analyzing performance of multilayered front matched and backed piezoelectric ceramic transducers', *Acoust. Imaging*, 1978, **8**, 395–416.

12. A. McNab, K. J. Kirk and A. Cochran, 'Ultrasonic transducers for high temperature applications', *IEE Proc. A*, 1998, **145**(5), 229–236.

13. K. Caidahl et al., 'New concept in echocardiography: harmonic imaging of tissue without use of contrast agent', *Lancet*, Oct. 17 1998, **352**(9136), 1264–1270.

Effect of Sol Infiltrations on the Electrical Properties of PZT

R. A. Dorey, R. D. Haigh, S. B. Stringfellow and
R. W. Whatmore

*School of Industrial and Manufacturing Science, Cranfield University, Cranfield, Bedfordshire
MK43 0AL, UK*

ABSTRACT

Thin PZT films with a thickness of approximately 1 µm can readily be fabricated using a layered sol–gel deposition technique. The maximum thickness obtainable is limited by the time taken and the tendency of the films to crack and spall when many layers are deposited. Thicker layers may be obtained by depositing a powder–sol slurry whereby a PZT powder is mixed with a sol of approximately the same composition. Using this layered slurry deposition technique, it is possible to obtain films with a thickness in excess of 20 µm. The resulting films, however, are often porous leading to poor properties and making subsequent patterning difficult.

A technique for increasing the density of such films through the use of controlled heat treatments and sol infiltration is presented. It is shown that with increased levels of sol infiltration the density and the dielectric constant of the films are maximised. Measurements of piezoelectric properties indicate that sol infiltrations have no significant effect on d_{33}. A sample with approximately 10% closed porosity was obtained following four infiltration steps per layer. This resulted in a maximum dielectric constant of approximately 700 and a d_{33} of 70 pC/N (poling conditions: 5 min at 200°C).

Examination of cross-sections of the films produced shows that for intermediate levels of sol infiltration (typically between 1 and 3) a density gradient is obtained with higher densities observed nearer to the base of the film. It is postulated that the observed density gradient is a result of continued infiltration from the upper layers when further layers of slurry are deposited.

The effects of changing process variables, such as the number of infiltration steps and firing temperature upon the film structures and properties, will be discussed.

INTRODUCTION

The maximum film thickness that can be achieved using a sol–gel processing route is often limited by the time taken and the tendency of the films to crack and spall. Thick films have been successfully produced[1–7] by spinning a composite slurry composed of PZT powder and a PZT producing sol. Typically, the relative permittivity of these films is considerably lower than that of bulk PZT. This difference in relative permittivity is due to the high level of porosity[1] within the film and the small size of the sol derived PZT crystallites.[7]

High levels of PZT powder loading have been shown to lead to high polarisation and coercive fields.[1] However, this also leads to a reduction in the relative permittivity due to the formation of high levels of porosity. Ohno et al.[1] demonstrated that a final sol infiltration step lead to an increase in the relative permittivity due to the reduction in the level of porosity.

The extent to which the liquid sol will penetrate into the porous film is limited by the pressure of the trapped gas within the pores counteracting the capillary pressure. Equation (1) gives the infiltration depth fraction (d_i) as a function of surface energy of the liquid/vapour interface ($\gamma_{l/v}$), pore radius (r) and contact angle (ϕ).

$$d_i = \left(1 - \frac{P r}{2\gamma_{l/v} \cos\phi}\right) \tag{1}$$

where P is atmospheric pressure.

For the system being examined, the pore radius is approximately 0.125 μm (~1/2 grain size), while the $\gamma_{l/v}\cos\phi$ term is likely to be in the range of 20–30 $\times 10^{-3}$ N m^{-1}. This gives rise to an infiltration efficiency of between 70 and 80%. Hence for a 10 μm thick film, approximately 2 μm of the material will not be infiltrated. This represents between four and five layers of PZT powder particles (mean grain size 0.5 μm). To further increase the infiltration depth an over pressure would have to be applied. An alternative approach, as used in this study, is to infiltrate each composite layer before the next layer is deposited. The limitations of infiltration still apply. However, the absolute thickness of non-infiltrated material is reduced to less than one particle layer. Further, the reduced probability of trapping gas and the wetting behaviour of the sol will favour complete infiltration.

The work undertaken in this study aimed to examine the variation in dielectric constant and piezoelectric properties with increased levels of sol infiltration. A high powder loading was selected to maximise the ferroelectric properties of the thick film. Repeated sol infiltration of the individual composite layers was then conducted in an attempt to maximise the dielectric properties of the film. The variation in the film properties was then recorded as a function of the number of sol infiltrations. To further aid densification of the thick film a liquid phase sintering aid[8] was added to the composite slurry.

EXPERIMENTAL PROCEDURE

Composite Slurry Production

A PZT sol of nominally identical composition to that of the PZT powder was produced using the route shown in Fig. 1. A composite slurry was then made by mixing the PZT sol with a PZT powder (Ferroperm PZ26) to produce a powder loading of 1.5 g mL; 2 wt% (relative to

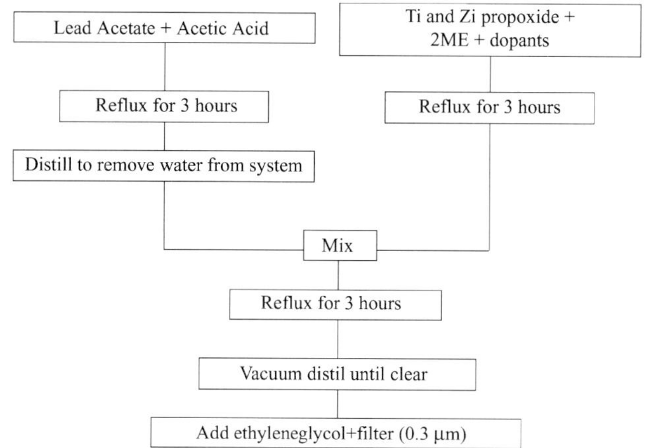

Fig. 1   PZT producing sol processing route.

the PZT powder mass) of a dispersant (KR55) was also added to ensure thorough dispersion of the PZT powder. Finally 4.7 wt% (relative to the PZT powder mass) of sintering aid $(0.2CuO_2–0.8PbO)$ was added to the slurry. Prior to use, the slurry was lightly ball milled for 24 h to ensure thorough mixing of all of the constituent parts.

Spin Coating and Infiltration

The PZT thick films were deposited onto platinised silicon wafers $(Pt/TiO/SiO_2/Si)$. Prior to coating the wafers were cleaned in acetone, followed by isopropanol and finally any residual organics were removed using a plasma ashing process in oxygen. The PZT thick films were built up by depositing a series of layers. Each layer consisted of an initial composite layer which was repeatedly infiltrated with sol. The composite layer was deposited by covering the entire wafer surface with the composite slurry and then spinning at 2000 rev min^{-1} for 30 s. The wafer was then subjected to a heat treatment designed to remove the organic component and crystallise the sol, resulting in a porous layer. Various numbers of repeat sol infiltrations were then conducted ensuring that each sol layer was subjected to the full firing regime before the next sol infiltration step was conducted. Once the required number of layers had been deposited, the wafer was subjected to a rapid thermal annealing (RTA) process designed to develop the perovskite structure. The nomenclature adopted during this study describes the deposition treatments such that:

(C2000-200, 450 + 2S2000-200, 450)4   710C30M

relates to a sample made up of four layers where each layer was composed of one composite layer spun at 2000 rev min^{-1} (C2000) and 2 sol layers each spun at 2000 rev min^{-1} (2S2000). Each deposition was subjected to a heat treatment of 200°C for 1 min and 450°C for 15 s. The final RTA treatment was at 710°C for 30 min (710C30M).

Electrical Measurements

Circular Cr–Au electrodes were evaporated onto the surface of the PZT films for electrical measurements. Contact was made with the back electrode by removing a small section of PZT by mechanical abrasion. Capacitance measurements between the back electrode and Cr–Au top electrodes were made using a component analyser (Wayne Kerr 6425B) at 1 kHz. The thickness of the films was determined from optical microscopy and SEM (ABT-55) observations of cross-sectional fracture surfaces. Values of thickness and capacitance were then used to calculate the relative permittivity.

The thick films were poled at 200°C for 5 min using a field of 8 V/μm. The piezoelectric coefficient d_{33} was measured using a Berlincourt piezometer.

RESULTS AND DISCUSSION

Relative Permittivity

Preliminary work examining sol infiltration employed a 450°C heating for the composite layer and a 200°C heat treatment for the sol infiltrate, as this had previously been used successfully for the preparation of sol–gel films.[9] However, this technique lead to a layered structure as shown in Fig. 2. The layered structure was attributed to the incomplete pyrolysis of the sol at 200°C which prevented further infiltration of fresh sol. Subsequent sol deposition stages lead to the thickening of the sol derived layers. It was therefore decided to employ both a 200 and 450°C heating stage to facilitate controlled and complete decomposition of the sol.

Fig. 2   SEM photomicrograph of a polished cross-section of a layered sol composite structure caused by incomplete pyrolysis of the sol.

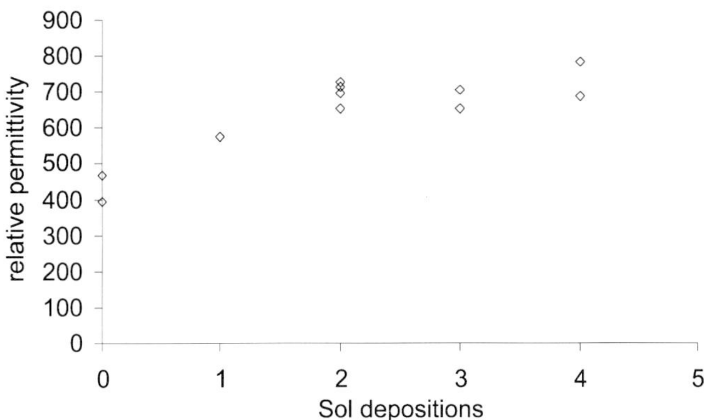

Fig. 3   Variation in relative permittivity with increased levels of sol infiltrtion (C2000-200, 450 + XS2000-200, 450)4   710C30M.

Figure 3 shows the relative permittivity of films produced using the (C2000-200, 450 + XS2000-200, 450)4 procedure where X (sol infiltration steps) was varied between 0 and 4. It can be seen that the relative permittivity of the film increases with increased levels of sol infiltration. The first two sol infiltrations result in the greatest increase in relative permittivity. Subsequent additions do not have as great an influence on relative permittivity. The increase in dielectric constant with increasing sol depositions indicates that the sol does infiltrate into the composite layer reducing the level of porosity. This is confirmed by observations of the fracture surface of thick films which shows that the level of porosity decreases with increased levels of sol infiltration (Fig. 4).

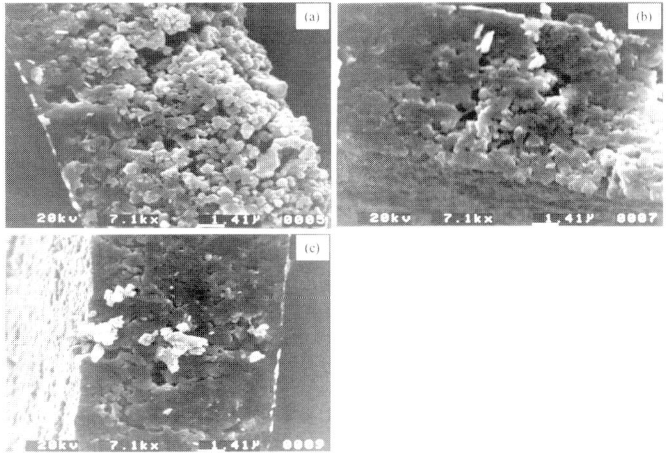

Fig. 4   Fracture surfaces of (C2000-200, 450 + XS2000-200, 450)4   710C30M films produced with the addition of 4.7 wt% sintering aid: (a) X=0; (b) X=2; (c) X=4.

The change in slope of the curve shown in Fig. 3 is thought to be due to the lower efficiency of the infiltration process when the pores are small and the development of a thin sol derived layer at very high infiltration levels (X=4). Despite the development of such a discrete layer at high infiltration levels, the dielectric constant is still increased. This is probably due to the beneficial effects of reducing the porosity which outweighs the detrimental effect of forming the layer. However, further sol depositions would be expected to result in a decrease in the dielectric constant as the sol derived layer thickens.

The sol infiltration technique employed in this study leads to a graded structure when low numbers of sol infiltrations are employed. This is due to the infiltration of sol from higher levels during the processing. Figure 4a clearly shows that the density is highest near the PZT/wafer interface. The infiltration of the sol into lower levels indicates that the interlayer strength should be relatively high as the sol forms a 3D network throughout the whole film. With the incorporation of three or four sol infiltration steps the variation in density is effectively removed and a homogenous structure is obtained. It can also be seen from Fig. 4 that there does not appear to be any regions where sol has not infiltrated indicating that that this technique is effective at ensuring full infiltration.

Piezoelectric Properties

Figure 5 shows the piezoelectric coefficient d_{33} as a function of sol infiltration steps. It can be seen that there is a large degree of scatter associated with the results. This is primarily due to the sensitivity of the measurement technique to the exact position of the probes relative to the electrode. There does not appear to be any significant variation in d_{33} with the number of sol infiltration stages. This is reasonable to expect as porosity is unlikely to constrain the PZT and hence effect the piezoelectric coefficient. A mean value of approximately 60 pC/N was observed, irrespective of the level of sol infiltration.

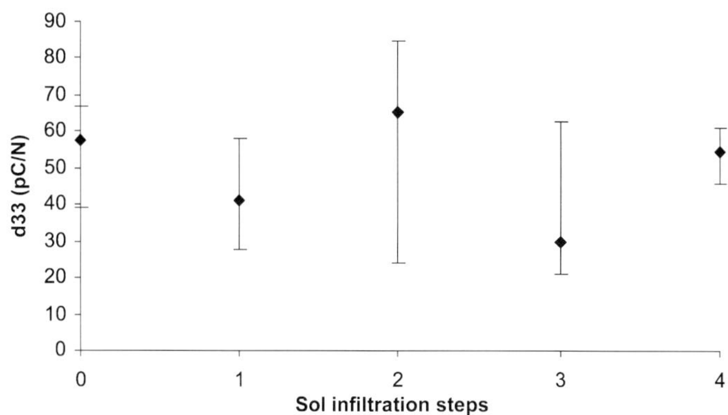

Fig. 5   Variation in d_{33} with the number of sol infiltration stages.

CONCLUSIONS

It has been shown that the incorporation of a series of sol infiltration stages gives rise to an increase in the relative permittivity. This is thought to be primarily due to the reduction in the level of porosity.

Complete infiltration of the sol can be achieved by infiltrating each composite layer prior to the deposition of the next layer.

The piezoelectric coefficient d_{33} does not exhibit any significant variation with the number of sol infiltration stages.

The use of intermediate heating stages of 200 and 450°C have been shown to be beneficial for ensuring the infiltration of the sol into the porous composite network. If the final 450°C heating stage is not used the sol will not fully pyrolise and the composite film will be effectively capped. Further sol depositions are then likely to lead to a layered structure. A layered structure consisting of composite layers with distinct dielectric constants could be produced if excessive sol depositions are avoided and the individual composite layers are infiltrated to different extents.

REFERENCES

1. T. Ohno, M. Kunieda, H. Suzuki and T. Hayashi, 'Low-temperature processing of $Pb(Zr_{0.53}Ti_{0.47})O_3$ thin films by sol–gel casting', *Jpn J. Appl. Phys.*, 2000, **39**, 5429–5433.
2. Y. Jeon, J. Chung and K. No, 'Fabrication of PZT thick films on silicon substrates for piezoelectric actuators', *J. Electroceram.*, 2000, **4**(1), 195–199.
3. D. A. Barrow, T. E. Petroff, R. P. Tandon and M. Sayer, 'Characterisation of thick lead zirconate titanate films fabricated using a new sol–gel beast process', *J. Appl. Phys.*, 1997, **81**(2), 876–881.
4. T. Tsurumi, S. Ozawa, G. Abe, N. Ohashi, S. Wada and M. Yamane, 'Preparation of $Pb(Zr_{0.53}Ti_{0.47})O_3$ thick films by an interfacial polymerisation method on silicon substrates and their electric and piezoelectric properties', *Jpn J. Appl. Phys.*, 2000, **39**, 5604–5608.
5. D. A. Barrow, T. E Petroff and M. Sayer, 'Thick ceramic coatings using a sol–gel based ceramic–ceramic 0–3 composite', *Surf. Coat. Technol.*, 1995, **76**(77), 113–118.
6. Q. F. Zhou, H. L. W. Chan and C. L. Choy, 'PZT ceramic/ceramic 0–3 nanocomposite films for ultrasonic transducer applications', *Thin Solid Films*, 2000, **375**, 95–99.
7. M. Lukacs, M. Sayer and S. Foster, 'High frequency ultrasonics using PZT sol–gel composites', *Intergrated Ferroelectrics*, 1999, **24**, 95–106.
8. D. L. Corker, R. W. Whatmore, E. Ringgaard and W. W. Wolny, 'Liquid-phase sintering of PZT ceramics', *J. Eur. Ceram. Soc.*, 2000, **20**, 2039–2045.
9. R. Haigh, 'Ferroelectric thick films prepared by chemical solution deposition of sol–gel composte slurries', MSc Thesis, Cranfield University, 2000.

Comparison of Piezoelectric Properties of a MOD and a Sol–Gel Niobium Doped PZT Thin Film

F. Duval, R. Dorey, S. Stringfellow, Q. Zhang and R. Whatmore

School of Industrial and Manufacturing Science, Cranfield University, Bedfordshire MK43 0AL, UK

ABSTRACT

Thin films of niobium doped lead zirconate titanate (PNZT) were prepared from a sol doped with 2 mol.% niobium and 10 mol.% lead excess. The sol possesses the specific formula $Pb_{1.1}Nb_{0.02}[Zr_{0.52}Ti_{0.48}]O_3$. Niobium being a soft dopant, it enhances the extrinsic contribution to the piezoelectric coefficient, and the dielectric properties. Films were deposited by spin coating and the dielectric properties assessed. Typically, 10 layers of sol were deposited by spinning onto platinum–titanium–silicon substrates followed by drying and annealing. Perovskite crystallisation was studied by XRD as a function of the sintering temperature from 550°C up to 740°C. Films were assessed by measuring capacitance and dielectric loss; d_{33} measurements were carried out after poling at 130°C for 5 min. The properties of the films were compared with those of films of a similar composition made using a commercial metallo-organic decomposition (MOD) solution. For the MOD process, the best properties were obtained with a 10 layer film and gave a dielectric constant of 650 for a thickness of 0.9 μm. A d_{33} of 62 pC/N was reached with a poling voltage of 13 V/μm. Sol–gel derived films exhibit a dielectric constant of 770 for a thickness of 1.1 μm and a d_{33} of 75 pC/N with a poling voltage of 15 V/μm.

INTRODUCTION

Sol–gel production of lead zirconate titanate (PZT) thin film has been investigated extensively over the last twenty years. The technique has the advantages of giving good control over the films chemical composition, including those with low levels of dopants. This paper deals with the preparation of niobium doped lead zirconate titanate (PNZT) thin films produced from sols doped with 2 mol.% of niobium. The addition of the dopant enhances the dielectric constant and piezoelectric coupling and reduces mechanical Q.[1] Different physical and dielectric properties have been obtained from sols produced using different precursors.[2–7]

Applications of PNZT include optical coating, thin capacitors for memory devices and also ultrasonic imaging devices for eye, skin or bones diagnostics.[8,9]

The company Protavic based in France has prepared a metallo-organic decomposition (MOD) solution.[10] Due to the degree of polymerisation within the MOD solution, a high viscosity is expected. The carbon content is also greater than sols obtained from classical alcohol routes due to the long carbon chain precursors used for the synthesis. Moreover, to evaporate all the organic compounds a higher pyrolysis temperature is usually required which may then activate large shrinkage during the drying process. The higher carbon content may also require a higher annealing temperature. A sol is of interest as it has a low viscosity compared with the MOD solution. The precursors employed were lead acetate trihydrate, titanium iso-propoxide, zirconium n-propoxide and niobium ethoxide. The sol used a very common solvent (ethanol) instead of the toxic 2-methoxyethanol. The objective of the study reported here was to compare the properties of films made via both routes.

EXPERIMENTAL PROCEDURE

Sol Synthesis

The A cation solution was prepared with 10 mol.% excess lead. Lead acetate trihydrate was placed into a round bottomed flask (250 mL) and dehydrated at 100°C under vacuum for 18 h; 15 mL of ethanol and 1.1 g of methylethyl alcohol (MEA) were then added to dissolve the lead acetate. A gentle heating was employed to dissolve the lead acetate completely. The B cation precursors employed were titanium iso-propoxide, zirconium n-propoxide and niobium ethoxide. All weighing was carried out in a glove box to prevent the stock solutions from being hydrolysed. The B cation solution was refluxed under nitrogen for 1 h. The A and B cation precursors were then mixed together and refluxed for a further 2 h in order to produce the PZT sol. A green solution was obtained. When cool, the sol was filtered and stabilised by the addition of 1 g of ethylene glycol. The final sol exhibited a pH of approximately 7, a density of 0.92 g cm^{-3}, a viscosity of 0.019 Poise and a molar concentration of 0.4 M. For comparison, the MOD solution exhibits a density of 1.01±0.01 g cm^{-3} and a viscosity of 25.25 Poise.

Film Deposition

Sol deposition for both MOD and sol–gel processes were performed in a clean room to avoid dust contamination. Si wafers with a Pt/Ti electrode were used as substrates in this study. Before deposition the substrates were cleaned with acetone and then with isopropanol, ensuring that a constant flow of solvent was maintained. Finally the samples were cleaned under an oxygen plasma by using a PT 7160 RF Plasma Barrel Etcher. The substrates were coated fully with the sol or MOD solution and then spun for 30 s. A large amount of liquid was deposited and the excess was rapidly spun off resulting in a uniform thin layer of sol. Intermediate firings were performed using a hot plate to remove volatile compounds. A final pyrolysis was then carried out to eliminate all the organic compounds. Typically 10 layers were deposited followed by a final annealing stage. The annealing step was conducted using a rapid thermal annealer (RTA) with a ramp of 30°C per second. The properties of the PZT films were investigated on samples fired using different annealing temperatures for 30 min. The films were then characterised electrically and crystallographically. The effects of spin speed, sol viscosity, intermediate firing treatments and final annealing were investigated.

Electrical Characterisation and X-Ray Diffraction

Gold chromium electrodes were deposited on the surface of the films by vacuum evaporation using a Edwards I Evaporator. The capacitance and dielectric loss were measured using an Wayne Kerr 6425 Analyser. To evaluate the capacitance a low voltage of 0.1 V and a frequency of 1 kHz were used. For each electrode, the capacitance value was taken as the average of three readings. Film thicknesses were evaluated by etching the corner of the samples to remove PZT. A Dektak® analyser was used to measure the height of the step between the bottom electrode and the top of the film. The samples were poled at 130°C for 5 min before measuring d_{33}; d_{33} values were measured using a Piezometer System PM 25. The poling voltage was increased stepwise to determine the d_{33} saturation values. X-ray diffraction was carried out to examine the evolution of the perovskite phase as a function of the annealing temperature. The X-ray diffraction was conducted using a Siemens diffractometer using a radiation Cu K_α. The conditions employed were a 2θ range from 10 to 50° with a step of 0.02° and a dwell time of 1 s.

RESULTS AND DISCUSSION

Deposition Technique

The conditions used to study the effect of the sintering temperature are summarised in Table 1. To obtain good films the effects of spin speed, viscosity of the sol and the firing treatments were investigated. It was found that wetting behaviour and uniformity of thickness could be

Table 1   Deposition condiions (1) for MOD sol and (2) for ethanol sol.

| | Spin coater | | Pre-bake | | | | Pyrolysis | | Annealing | |
| | | | 1 | | 2 | | | | | |
	Speed (rev min^{-1})	Time (s)	Temp. (°C)	Time (s)	Temp. (°C)	Time (s)	Temp. (°C)	Time (s)	Temp. (°C)	Time (s)
(1)	3000	30	200	60	300	60	530	15	550–740	1800
(2)	2000	30	200	30		450 or 530	15 to 600	450 or 530	600–740	1800

Table 2   Relative intensities of XRD peaks for PZT films annealed at different temperatures for the MOD solution.

Annealing temperature (°C)	(100)	(111)	(200)
550	60	3	37
580	61	4	35
610	60	4	36
670	60	6	34
740	60	3	37

Table 3   Relative intensities of XRD peaks for PZT films annealed at defferent temperatures for the ethanol sol (* different pyrolysis conditions were applied to these films).

Annealing temperature (°C)	(100)	(111)	(200)
600	23	68	9
650	25	67	8
680	29	64	7
710*	25	68	7
710*	28	64	8
740*	30	63	7

improved by using a spin speed of 3000 rev min^{-1} and a diluted MOD solution (1:2 solution:ethanol dilution). In order to reduce the shrinkage of the film a second pre-bake at 300°C for 1 min was added to the procedure to remove high-chained alcohol molecules used as precursors. Finally, the pyrolysis temperature was increased from 450 to 530°C, so that no further weight was lost (data provided by a TGA analysis from Protavic).

X-Ray Diffraction

The films do not exhibit a preferred orientation as can be seen by the relative intensities quoted in Tables 2 and 3. The relative intensities were normalised with respect to the (100) perovskite peak at 22°.

For annealing temperatures below 610°C for the MOD solution and 680°C for the ethanol sol, the pyrochlore phase is still present (Figs 1 and 2). The two most important perovskite orientations are (100) and (200) for the MOD process; also the films exhibit a preferred orientation whilst it is mainly the (100) and the (111) orientations which are present in the sol–gel derived films. This suggests coexistence of the rhombohedral and tetragonal phases.

Alonso et al.[11] reported that the pyrochlore phase was detectable up to a temperature of 600°C for undoped PZT 52/48 thin films. Lead excess aids the development of the perovskite phase at lower temperatures. One of the reasons to explain the random orientation of the films might be the rapid temperatures ramp rate.[12]

Dielectric and Piezoelectric Properties

MOD process

Table 4 (p) and the Figs 3 and 4 summarise the electrical results obtained for the MOD produced films. An increase in the annealing temperature leads to an increase in the dielectric constant. At 740°C the dielectric constant has a maximum value of 650. The dielectric loss measured before poling was between 5 and 6.5%. The relative large value for the loss may suggest the presence of a residual amorphous phase due to an excess of lead oxide.[12]

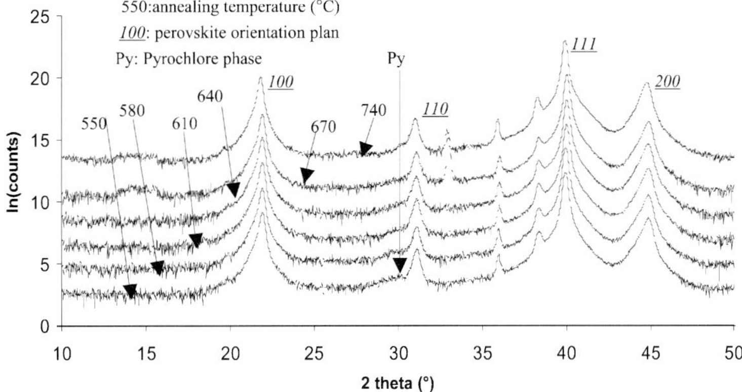

Fig. 1   XRD patterns of intensity versus 2θ for PZT films annealed at different temperatures from the MOD process.

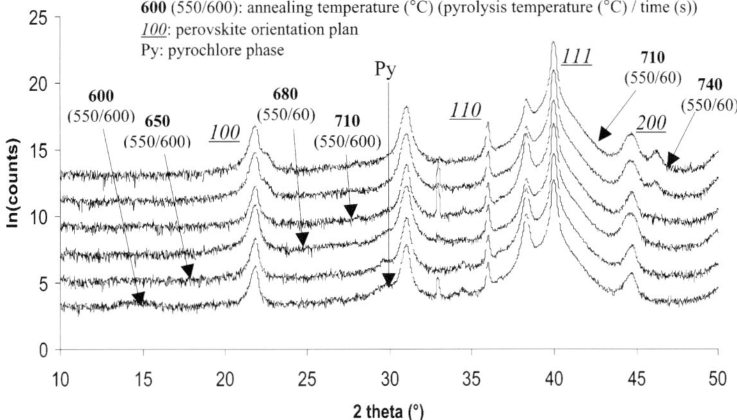

Fig. 2   XRD patterns of intensity versus 2θ for PZT films annealed at different temperatures from the ethanol-based sol.

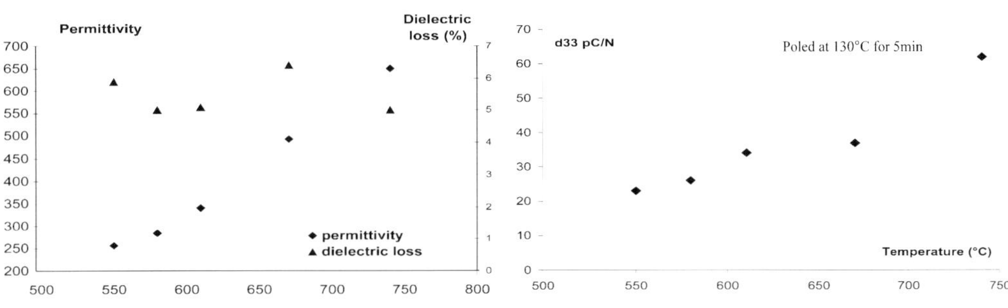

Fig. 3   Relative permittivity and dielectric loss versus the sintering temperature for the MOD process.

Fig. 4 d_{33} versus the sintering temperature (values of d_{33} were obtained using a saturation poling field).

Table 4   Piezoelectric properties obtained from the MOD process for different annealing temperatures.

RTA Temperature (°C)	Thickness (µm)	Dielectric loss (%)	Dielectric constant	d_{33} (pC/N)
550	0.6	5.9	250	23
580	0.6	5.0	280	26
610	0.6	5.1	340	34
670	0.6	6.4	500	37
740	0.9	5.0	650	62

The value of d_{33} increases with the temperature with a maximum of 62 pC/N obtained at 740°C.

Sol–gel process

Table 5 summarises the dielectric properties obtained from the sol–gel process. The films obtained from the sol–gel process exhibit the same thickness of 1 mm after 10 layers were deposited. This is to be expected as the only process variable employed was the annealing temperature, which should only affect the phases present. In fact, the viscosities of both diluted MOD solution and ethanol sol were found to be similar after measuring with a Bohlin® Instrument rheometer at 200 s⁻¹. The dielectric constant was not observed to increase with the RTA temperature. The best properties in terms of permittivity and d_{33} were obtained for a pyrolysis temperature at 550°C for 1 min and annealing at 680°C for 30 min.

A low temperature, 450°C, activates cracks very quickly inside the film. Increasing the pyrolysis temperature to 530°C enabled organic compounds to be eliminated without activating cracking during the process. A longer process, 10 min instead of 1 min, at the same temperature reduces the dielectric properties; nevertheless the short process of 15 s used for the MOD derived films seemed to cause a very fast organic evaporation responsible for the defects observed.

As reported in the literature,[13,14] an optimum drying temperature at which the film exhibits the finest microstructure enables the highest relative permittivity to be obtained. In fact, the pyrochlore phase ratio can increase again when the temperature goes up.

Miyazawa[16] reported that increasing the time of pyrolysis degraded the electrical properties of PZT (52/48) thin films. The dielectric constant obtained for a pyrolysis step at 550°C

Table 5   Piezoelectric properties obtained from the sol–gel process for different annealing temperatures (+ data impossible to measure because the electrode were damaged).

Pyrolysis Temp.(°C) (time (s))	RTA Temp. (°C)	RTA Time (s)	Thickness (µm)	Dielectric loss (%)	Dielectric constant	d_{33} (pC/N)
550 (600)	710	1800	1.05	2.6	520	40
550 (60)	680	1800	1.14	3.6	770	75
550 (60)	710	1800	0.91	+	585	+
550 (60)	740	1800	1.03	+	665	73

for 10 min is only about 500 compared with a value of 600 for a pyrolysis at the same temperature for 1 min.

CONCLUSIONS

Niobium doped PZT thin films were prepared using MOD and sol–gel routes. For the successful production of crack free MOD derived thin films, it was necessary to reduce the viscosity of the solution by diluting with ethanol. This increased the spreading speed of the solution and the solvent evaporation rate. Annealing temperatures in excess of 610°C were found to be necessary for the removal of the pyrochlore phase for MOD derived films. It was shown that increases in annealing temperature, over the range of 550–740°C, lead to an increase in the dielectric constant and d_{33}. Maximum values of 650 for dielectric constant, and 62 pC/N for d_{33}, were obtained using an annealing temperature of 740°C. The minimum pyrolysis temperature for the production of crack free sol–gel derived thin film was found to be 550°C. An annealing temperature greater than 710°C was shown to be necessary to eliminate the pyrochlore phase. The maximum dielectric constant was obtained for samples sintered at 680°C. Further increase in temperature resulted in a decrease in dielectric constant. d_{33} was found not to change with annealing temperature. Maximum values of dielectric constant and d_{33} were 770 and 75 pC/N, respectively.

ACKNOWLEDGEMENTS

The authors would like to thank Protavic for supplying the MOD solution. The financial support of EPSRC through project GR/N05970 and the CEC through the PARMENIDE project is gratefully acknowledged.

REFERENCES

1. M. Dvorsek and M. Kosec, 'Microstructural and electromechanical properties of donor doped PZT Ceramics', *Sci. Ceram.*, 1988, **14**, 951–956.
2. Y. T. Kwon, I. M. Lee, W. I. Lee, C. J. Kim and I. K. Yoo, 'Effect of sol–gel precursors on the grain structure of PZT thin films', *Mater. Res. Bull.*, 1999, **34**(5), 749–760.
3. K. D. Budd, S. K. Dey and D. A. Payne, 'Sol–gel processing of PbTiO$_3$, PbZrO$_3$, PZT and PLZT thin films', *Brit. Ceram. Proc.*, 1985, **36**, 107–122
4. M. Sayer, G. Yi and M. Sedlar, 'Comparative sol gel-processing of PZT thin films', *Integrated Ferroelectrics*, 1995, **7**, 247–258.
5. R. W. Scharwtz, T. J. Boyle, S. J. Lockwood, M. B. Sinclair, D. Dimos and C. D. Buchheit, 'Sol–gel processing of PZT thin films: a review of the state of the art and process optimization strate-

gies', *Integrated Ferroelectrics*, 1995, **7**, 259–277.

6. R. W. Scharwtz, J. A. Voight, B. A. Tuttle, D. A. Payne, T. L. Reichert and R. S. DaSalla, 'Comments on the effects of solution precursor characteristics and thermal processing conditions on the crystallisation behavior of sol–gel derived lead zirconate titanate thin films', *J. Mater. Res.*, 1997, **12**(2), 444–456.

7. M. Klee, R. Eusemann, R. Waser, W. Brand and H. van Hal, 'Processing and electrical properties of $Pb(Zr_xTi_{1-x})O_3$ (x=0.2–0.75) films: comparison of metallo-organic deposition and sol–gel processes', *J. Appl. Phys.*, 1992, **72**(4), 1566, 1576.

8. M. Pudenziati (ed.), 'Thick film sensors', *Handbook of Sensors and Actuators*, 1994, 1.

9. 'Piezoelectric array for medical imaging and non destructive control using integrated micro electro mechanical devices', http://www.lcr.thomson-csf.com/projects/www_parmenide/index.html (12/12/2000)

10. European patent No. 89-15174. 'Procédé de dépôt d'une composition céramique en couche mince et produit obtenu par ce procédé'.

11. R. E. Alonso, P. de la Pressa, A. Ayala, A. Lopez-Garcia and C. Livage, 'The stability of $Pb(Zr_{0.52}Ti_{0.48})O_3$ prepared by the sol–gel method', *Solid State Communication*, 1998, **107**(4), 183–187.

12. H. Suzuki, T. Koizumi, Y. Kondo and S. Kanako, 'Low-temperature processing of $Pb(Zr_{0.53}Ti_{0.47})O_3$ thin film from stable precursor sol', *J. Eur. Ceram. Soc.*, 1999, **19**, 1397–1401.

13. C. Lee, V. Spirin, H. Song and K. No, 'Drying temperature effects on microstructure, electrical properties and electro-optic coefficients of sol–gel derived PZT thin films', *Thin Solid Films*, 1999, **340**, 242–249.

14. C. W. Law, K.Y. Tong, J. H. Li and K. Li, 'Effect of pyrolysis temperature on the characteristics of PZT films deposited by the sol–gel method', *Thin Solid Films*, 1998, **335**, 220–224.

15. W. I. Lee and J. K. Lee, 'Dopants effects on the grain structure and electrical property of PZT thin films prepared by sol–gel process', *Mater. Res. Bull.*, 1995, **30**(10), 1185–1191.

16. K. Miyazawa, K. Ito, and R. Maeda, 'Structure and electrical properties of multilayer PZT films prepared by sol–gel processing', *Ceram. Int.*, 2000, **26**, 501–506.

Functional Behaviour of Thin Film Dielectric Superlattices

J. M. Gregg,* M. H. Corbett, D. O'Neill, G. Catalan and R. M. Bowman

Department of Pure and Applied Physics, The Queen's University of Belfast, Belfast BT7 1NN, UK

ABSTRACT

Pulsed laser deposition has been used to fabricate thin film capacitor structures in which the dielectric layer is a superlattice. The properties of two superlattice systems were investigated as a function of superlattice wavelength (Λ) – one based on barium strontium titanate and the other on lead-based relaxor electroceramics. In both systems the dielectric constant was significantly enhanced at stacking wavelengths of a few unit cells. However, the dielectric enhancement seen in the barium strontium titanate superlattices was found to be due to Maxwell–Wagner effects, whereas in the relaxor superlattices Maxwell–Wagner behaviour was not evident; rather, the dielectric enhancement was associated with the onset of polar coupling around $\Lambda \sim 20$ nm.

INTRODUCTION

Significant interest in dielectric superlattices has developed over the last decade, fuelled by the possibility of functional properties in superlattices being superior to those of compositionally equivalent solid solutions. Experimentally, dielectric constants have frequently been observed to increase on decreasing superlattice wavelength (Λ),[1–7] with the most dramatic study claiming a relative dielectric constant of 420,000. Other effects include reduced temperature dependence of dielectric properties and potentially dramatically enhanced polarisation.[5] Interesting physics has also emerged, with interlayer coupling occurring at relatively small Λ. Such results have encouraged recent modelling of dielectric superlattices in which coupling is considered.[8,9]

Unfortunately, there are serious inconsistencies in the body of research produced to date. The extent of dielectric enhancement on decreasing Λ varies dramatically, and is found to be a maximum at very different scales of heterostructure. Crucially, dielectric losses are not fully reported, or are high in much of the work. In an attempt to rationalise such inconsistencies

---

*Corresponding author's email: m.gregg@qub.ac.uk.

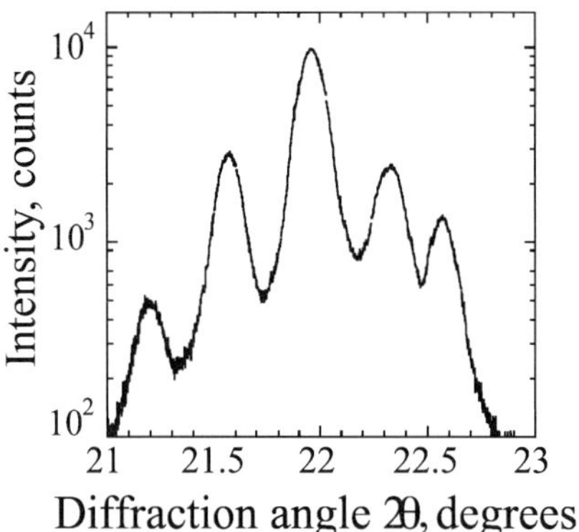

Fig. 1 θ-2θ X-ray diffraction shows the central Bragg peak with superlattice satellites.

the authors here report studies examining the properties of two superlattice systems: $Ba_{0.2}Sr_{0.8}TiO_3/Ba_{0.8}Sr_{0.2}TiO_3$ and $Pb(Mg_{1/3}Nb_{2/3})O_3/[0.2Pb(Zn_{1/3}Nb_{2/3})O_3–0.8BaTiO_3]$.

EXPERIMENTAL DETAILS

Thin film capacitors were made by pulsed laser deposition (PLD) as follows: $SrRuO_3$ lower electrodes were deposited onto single crystal {100} MgO substrates under 0.15 mbar of oxygen with a substrate temperature of 775–800°C. Superlattices were then deposited (BST-based at 775°C, and relaxor-based at 630°C), before post-deposition annealing under 900 mbar O_2. Specimens were then removed from the PLD system and two gold electrodes (~2 mm²) were deposited by thermal evaporation through a hard mask. Dielectric testing was performed making contact to two upper electrodes (two capacitors in series) and using Hewlett Packard LCR meters in conjunction with Oxford Instruments cryostats. Polarisation loops were measured using a Radiant Technologies RT6000 Precision Workstation. Structural characterisation was performed using a Siemens D5000 X-ray diffractometer (XRD), Brüker-AXS XRD and Tecnai F20 high-resolution transmission electron microscope (HRTEM) with energy dispersive X-ray (EDX) analysis and a high angle annular dark field detector (HAADF).

RESULTS AND DISCUSSION

Successful growth of superlattice structures was established by the observation of satellite peaks in XRD (Fig. 1), and by direct imaging under cross-sectional transmission electron microscopy (Fig. 2). The superlattice wavelengths were calculated from the XRD traces, and verified using HRTEM. The chemically distinct nature of the individual layers within the superlattice was also confirmed by HRTEM (Fig. 2).

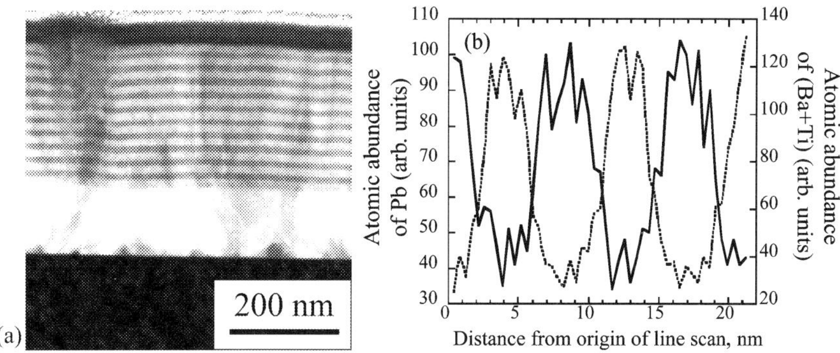

Fig. 2   Cross-sectional HRTEM investigation helped to verify superlattice wavelength and showed that the individual layers within the superlattice had remained chemically distinct. (a) HAADF image of a BST superlattice; (b) EDX line scan taken perpendicular to the interlayer interface in a lead based relaxor superlattice.

The detailed dielectric behaviour of the two series of superlattices investigated was found to be very different. For clarity, therefore, the results from the barium strontium titanate and relaxor superlattices will be presented and discussed separately below.

Barium strontium titanate (BST) superlattices

The dielectric behaviour of these superlattices as a function of individual BST layer thickness is shown in Fig. 3. The form of the response is extremely similar to that published by Tabata et al. for $BaTiO_3/SrTiO_3$ superlattices.[2,4] In their work, a maximum in dielectric

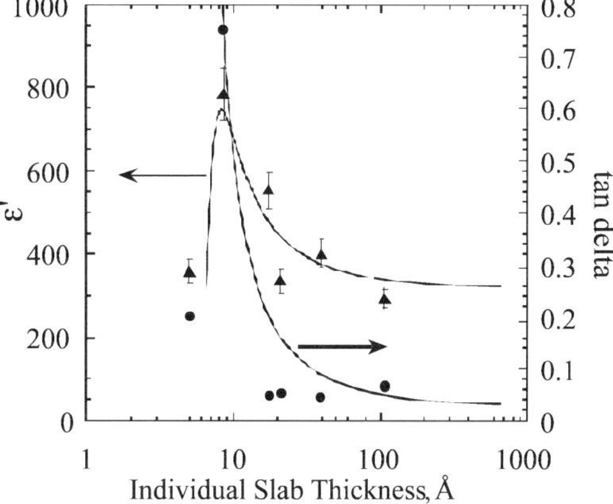

Fig. 3   Dielectric constant and loss at room temperature and 10 kHz for BST superlattices as a function of individual BST layer thickness. Solid line represents fit from MW model.

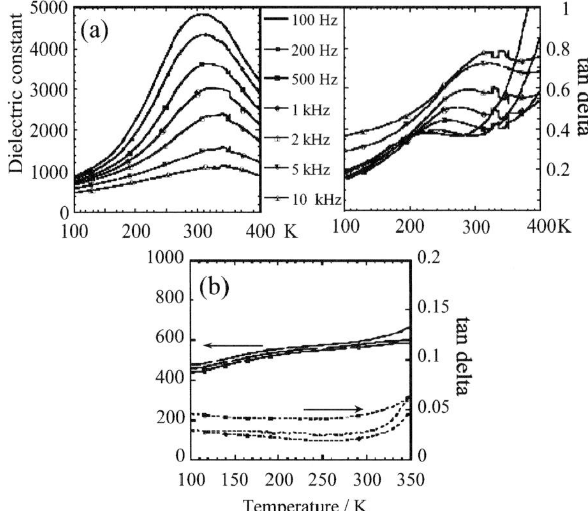

Fig. 4   Dielectric behaviour of superlattices with indiidual BST layer thicknesses of (a) 8 Å and (b) 16 Å as a function of temperature and frequency.

constant was also observed at individual dielectric layer thicknesses of 8–12 Å, and the background dielectric constant value of ~500 was enhanced by almost a factor of two to ~900 (compared with ~350 to ~800 in Fig. 3). Losses for individual layer thicknesses greater than 12 Å were also seen to be tan δ~0.05 as in our work. Losses for the fine multilayer structures associated with the dielectric peak were not presented by Tabata et al.

Figure 4 shows the dielectric behaviour of the 8 Å (Fig. 8a) and 16 Å (Fig. 8b) superlattices as a function of frequency and temperature. A change from the relatively invariant behaviour typical of ferroelectric thin films in coarser superlattices, to behaviour strongly reminiscent of relaxors in fine superlattices can be seen. Fine-scale superlattices also demonstrate frequency behaviour that is qualitatively similar to that presented by Erbil et al. for $PbTiO_3$–$Pb_{1-x}La_xTiO_3$ heterostructures,[1] with ε' increasing rapidly as frequency decreases. Such low frequency behaviour suggests that conduction mechanisms might play an important role. Results were therefore analysed in terms of the Maxwell–Wagner (MW) capacitor model.[10] It can be shown that the MW capacitor yields real and imaginary parts of relative permittivity.

$$\varepsilon'(\omega) = \frac{1}{C_0(R_1 + R_2)} \frac{\tau_1 + \tau_2 - \tau + \omega^2\tau_1\tau_2\tau}{1 + \omega^2\tau^2} \tag{1}$$

$$\varepsilon''(\omega) = \frac{1}{\omega C_0(R_1 + R_2)} \frac{1 - \omega^2\tau_1\tau_2 + \omega^2\tau(\tau_1 + \tau_2)}{1 + \omega^2\tau^2} \tag{2}$$

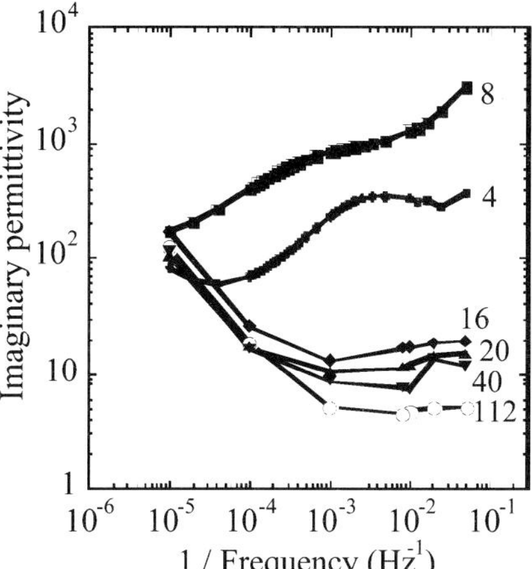

Fig. 5   Frequency dependence of the imaginary permittivity for BST superlattices (number indices refer to individual layer thicknesses in Å).

where $\tau_1 = C_1 R_1$, $\tau_2 = C_2 R_2$, $\tau = \dfrac{\tau_1 R_2 + \tau_2 R_1}{R_1 + R_2}$, $C_0 = \varepsilon_0 \dfrac{A}{t}$, A = capacitor area, t = thickness; ω = frequency; subscripts 1 and 2 refer to superlattice components. By considering behaviour at zero and infinite frequency, equivalent expressions to equations (1) and (2) can be given in terms of ε_0 and ε_∞

$$\varepsilon'(\omega) = \varepsilon_\infty + \frac{\varepsilon_0 - \varepsilon_\infty}{1 + \omega^2 \tau^2} \tag{3}$$

and

$$\varepsilon''(\omega) = \frac{1}{\omega C_0 (R_1 + R_2)} + \frac{(\varepsilon_0 - \varepsilon_\infty)\omega\tau}{1 + \omega^2 \tau^2} \tag{4}$$

Equation (3) is the same as that for Debye relaxation. However, the imaginary permittivity distinguishes between Debye and MW behaviour. In particular $\varepsilon'' \to 0$ as $\omega \to 0$ in a Debye system, whereas in a MW system $\varepsilon'' \to \infty$. Figure 5 shows ε'' for superlattices as a function of frequency. Clearly, the behaviour of superlattices composed of individual BST layers greater than 8 Å is significantly different from those of 8 and 4 Å. These finer superlattice structures show a tendency for ε'' to increase with frequency decrease, indicating the onset of the MW effect. Figure 6 demonstrates that the frequency response for the finest scale superlattices can be directly modelled by a MW expression and that a Debye expression cannot account for the low frequency behaviour.

The origin of the change in functional behaviour as individual layer thickness is reduced in the superlattice structures is not entirely clear. However, some insight was gained whenever the behaviour of the superlattices described above was compared with those in which

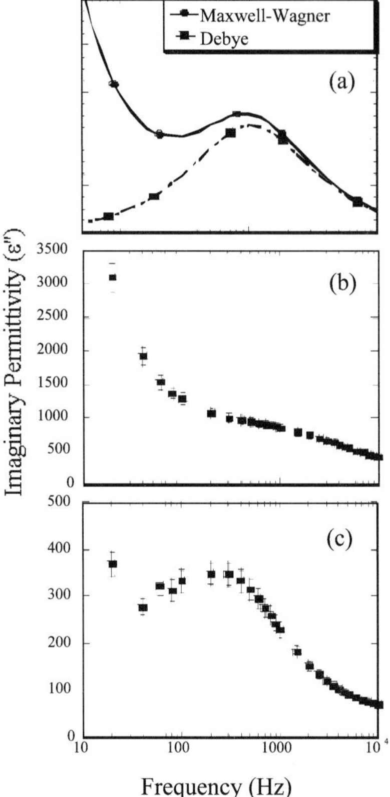

Fig. 6   Form of the imaginary permittivity response expected from (a) MW and Debye models, compared with that found in superlattices with individual BST layer thicknesses of (b) 8Å and (c) 4 Å.

delays were deliberately introduced between deposition of each successive layer in the superlattice stack. Figure 7 shows such a capacitor with individual layer thickness of 16 Å. Clearly, dielectric enhancement and frequency relaxation have been introduced. The lack of such features in the 16 Å superlattice without delays (Fig. 4) shows that absolute thickness of individual layers does not directly determine functional properties.

The top surfaces in individual layers in a capacitor with delayed deposition have been exposed to low pressures and high temperatures for long times in comparison to surfaces in superlattices without delays. Under such conditions surface modifications can be expected, their extent being dependent on the exposure time. In a given superlattice, the total extent of modification will also be proportional to the number of interfaces present. Hence, similar relative levels of surface-related modification in a superlattice capacitor can be achieved by either increasing exposure times or decreasing slab thickness. This is exemplified by the functional similarity between the 16 Å slab thickness capacitor with delays (Fig. 7) and the 8 Å slab thickness capacitor without delays (Fig. 4a).

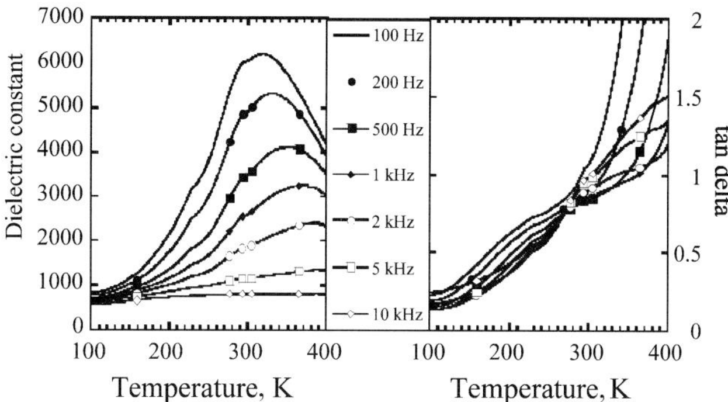

Fig. 7   The dielectric constant and loss for a 16 Å slab thickness BST superlattice with delays introduced between the deposition of each layer.

Increased permittivities, and associated frequency relaxation features, were consistently found to be correlated with increased conductivities. Surface modifications must therefore be such that they reduce the resistivity of the interfacial material, resulting in an overall decrease of resistance across the capacitor. We can again invoke the MW series capacitor treatment given above, but with the effective superlattice structure now consisting of interfacial, semiconducting regions intercalated between bulk-like, insulating ferroelectric.

The details of this modelling can be found elsewhere,[11] but it can clearly be seen from Fig. 3 that it is capable of reproducing the experimental observations of dielectric constant and loss as a function of superlattice wavelength. Moreover, when thermal and frequency dependence of the functional properties of the 'semiconductors' within the MW stack are considered, the full functional behaviour of the fine-scale BST superlattices can be reproduced (Fig. 8).

A reasonably convincing case can therefore be made for blaming features, such as enhancement in dielectric constant and 'relaxor-like' properties in fine-scale superlattices of BST, on Maxwell–Wagner effects caused by dielectric interface-related defects. However, as seen below, such an explanation is simply not appropriate for the behaviour of lead-based relaxor superlattices.

Lead-based relaxor superlattices

Dielectrically, all superlattices showed relaxor behaviour with a single value of T_m (Fig. 9). Frequency dispersion was significant below T_m, and negligible above T_m. In coarse superlattices, the dielectric constant was ~350–400 and was found to be relatively insensitive to changes in Λ. However, below $\Lambda = 20$ nm, the dielectric constant was found to steadily increase as Λ decreased (Fig. 10). In the finest-scale superlattices examined, dielectric constants were around a factor of 2–3 times greater than in coarse superlattices, and were

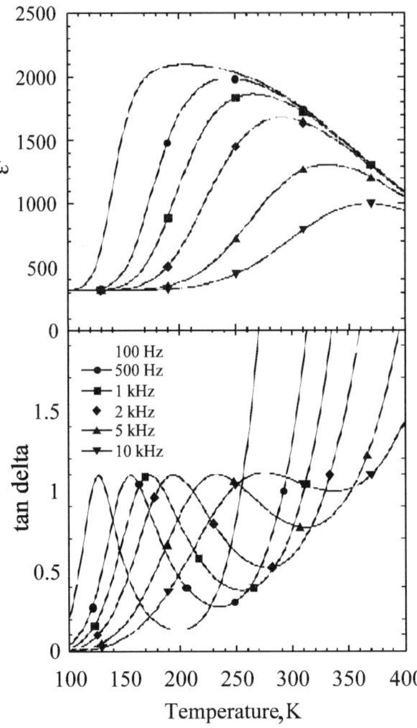

Fig. 8   Dielectric constant and loss as a function of temperature and frequency predicted by the model for a capacitor with t_i/t. Note the similarity with the capacitor behaviour in Figs 4 and 7.

also significantly greater than that of the equivalent solid-solution dielectric films – results very similar to those seen in the BST superlattices.

Importantly, the observed enhancement in dielectric constant was not associated with increases in loss tangent. There is therefore no suggestion that increased conductivity in the

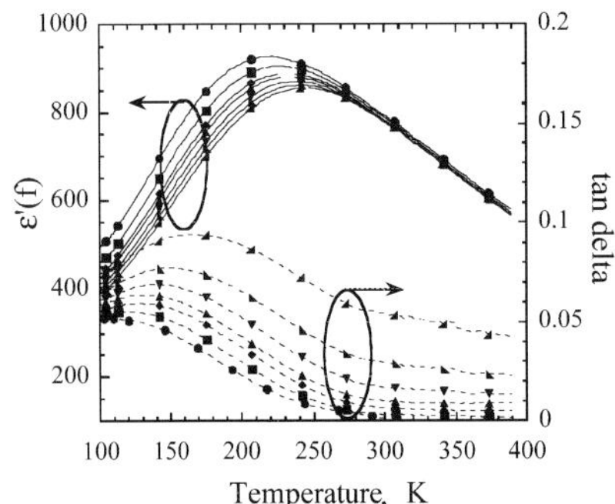

Fig. 9   Dielectric constant and loss tangent of lead-based relaxor superlattice (Λ=10 nm) as a function of temperature and frequency.

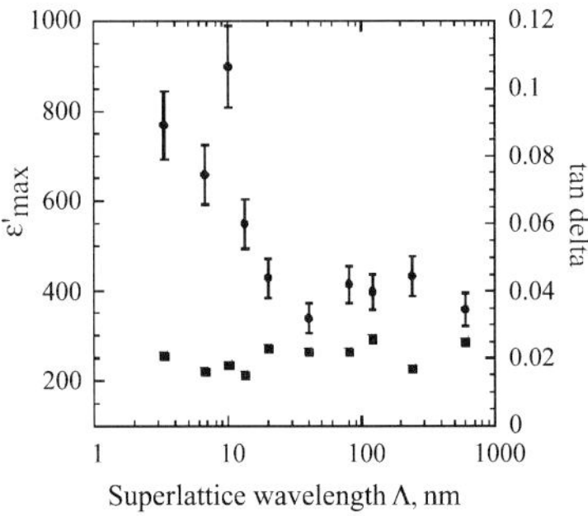

Fig. 10   Behaviour of the dielectric contant (circles) and loss tangent (squares) as a function of superlattice wavelength.

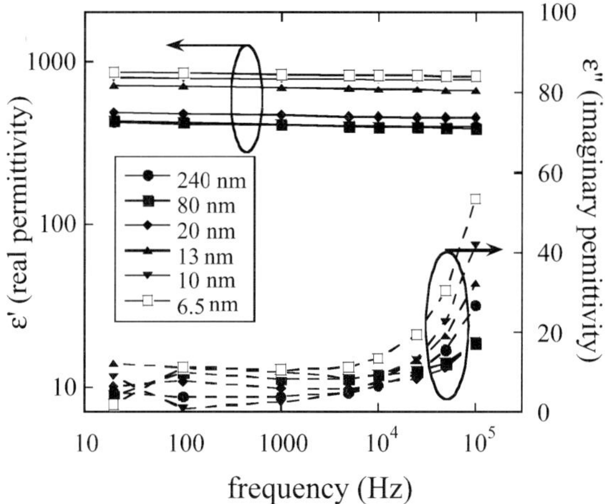

Fig. 11   Real and imaginary permittvity for superlattices at 300 K as a function of frequency.

fine-scale superlattices is responsible for the observed permittivity enhancement. This is reiterated in Fig. 11 where the imaginary permittivity as a function of frequency is explicitly demonstrated for the superlattice series. As can be seen, $\varepsilon'' \rightarrow 0$ as $\omega \rightarrow 0$, indicative of Debye-like rather than Maxwell–Wagner behaviour. The measured dielectric constant therefore seems to be genuinely associated with orientational polarisation in the superlattice stack.

It is tempting to interpret results in terms of recent Ising-spin modelling of superlattice electroceramic structures,[8,9] where distinct peaks in dielectric constant can be generated as a result of size-induced depolarisation on decreasing Λ. Indeed our results are extremely similar to those seen in Qu et al.[8] for uncoupled systems. However, this work ignores the fact that dielectric anomalies are not experimentally observed in thin film ferroelectrics on depolarisation (either induced by increased temperature or decreased size). Further, in the relaxor superlattice system investigated here, we find no evidence for size-induced depolarisation. Rather, polarisation is maintained across the series without any obvious trends.

It was hoped that studying the polarisation as a function of temperature might give insight into the freezing dynamics of the dual-relaxor system within the superlattice structure, and help rationalise the observed enhancement in dielectric constant. Polarisation loops were therefore taken as a function of temperature between 100 and 250 K, heating at 0.5 K min^{-1}. Even at 250 K (significantly above T_m for low frequency measurements) a remnant polarisation is apparent. Remnant polarisation above T_m seems to contravene established relaxor physics; nevertheless recent reports confirm this effect in relaxor thin films.[12,13]

In single crystal relaxor studies, anomalies in polarisation and associated linear birefringence have been observed and related to the transition from superparaelectric behaviour to the low temperature non-ergodic state.[14] Analysis of the changes in remnant polarisation with temperature for the relaxor superlattices (Fig. 12) showed that for large Λ (> 20nm) two anomalies in $-\dfrac{\delta P_r}{\delta T}$ can be seen. It is suspected that these correspond to polar melting events, and in

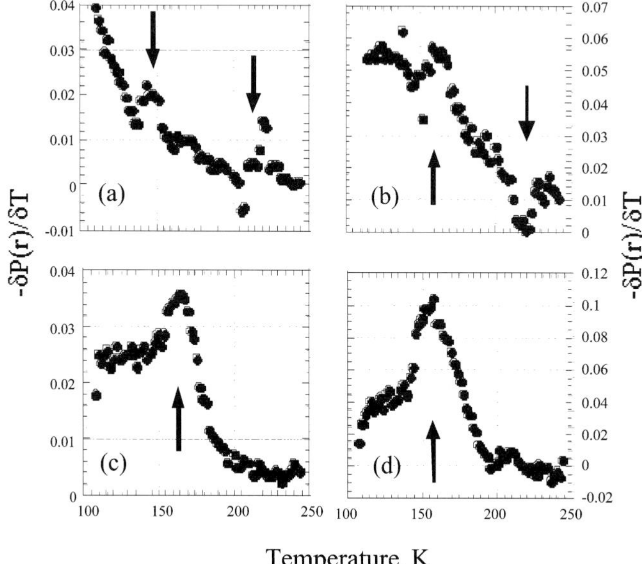

Fig. 12 $-\dfrac{\delta P_r}{\delta T}$ against T for superlattices with $L=$ (a) 600 nm, (b) 240 nm, (c) 20 nm and (d) 4 nm. Coare superlattices display two anomalies, while fine-scale superlattices shw only one anomaly.

this respect the anomaly at ~215 K seen in relatively coarse superlattices (Fig. 12a and b) is consistent with previous evaluations of T_f in single crystal PMN.[15,16] Below $\Lambda = 20$ nm, the double anomaly behaviour changes to a single dominant anomaly. Vögel–Fulcher modelling of the migration of T_m as a function of frequency for superlattices with $\Lambda \leq 20$ nm yielded freezing temperatures reasonably consistent with the single anomalies seen in $-\dfrac{\delta P_r}{\delta T}$, further validating their interpretation in terms of polar melting.

The transformation from a double to single polar melting event on decreasing Λ implies loss of independent functional behaviour in the two relaxor components of the superlattice structure. The length scales at which this coupling occurs is reasonably consistent with Specht et al. and their observations of structural coupling in the $KTaO_3/KNbO_3$ superlattice system[17] (structural phase transformation behaviour became strongly coupled at wavelengths between $\Lambda = 5$ nm and $\Lambda = 10$ nm). Since the onset of the single anomaly, and loss of double anomaly, is also associated with the point at which progressive enhancement in the dielectric constant occurs on decreasing Λ, results suggest that coupling of the polarisation dynamics is somehow responsible for the changes in dielectric constant observed.

CONCLUSIONS

The functional properties of two series of superlattices (one based on barium strontium titanate and the other on lead-based relaxors) have been investigated. In both cases an enhancement in dielectric constant was seen on decreasing superlattice wavelength. However, the origin of the enhancement was different: a Maxwell–Wagner effect in barium strontium titanate-based superlattices, and a polar coupling effect in relaxor-based superlattices.

REFERENCES

1. A. Erbil, Y. Kim and R. A. Gerhardt, *Phys. Rev. Lett.*, 1996, **77**, 1628.
2. H. Tabata, H. Tanaka, T. Kawai and M. Okuyama, *Jpn. J. Appl. Phys.*, 1995, **34**, 544.
3. I. Kanno, S. Hayashi, R. Takayama and T. Hirao, *Appl. Phys. Lett.*, 1996, **68**, 328.
4. H. Tabata, H. Tanaka and T. Kawai, *Appl. Phys. Lett.*, 1994, **65**, 1970.
5. O. Nakagawara, T. Shimutu, T. Makino, S. Arai, H. Tabata and T. Kawai, *Appl. Phys. Lett.*, 2000, **77**, 3257.
6. D. O'Neill, R. M. Bowman and J. M. Gregg, *Appl. Phys. Lett.*, 2000, **77**, 1520.
7. B. D. Qu, M. Evstigneev, D. J. Johnson and R. H. Prince, *Appl. Phys. Lett.*, 1998, **72**, 1394.
8. B. D. Qu, W. L. Zhong and R. H. Prince, *Phys. Rev. B*, 1997, **55**, 11218.
9. J. Shen and Y. Ma, *Phys. Rev. B*, 2000, **61**, 14279.
10. A. von Hippel, *Dielectrics and Waves*, Artech House, 1995.
11. G. Catalan, D. O'Neill, R. M. Bowman and J. M. Gregg, *Appl. Phys. Lett.*, 2000, **77**, 3078.
12. Z. Kighelman, D. Damjanovic and N. Setter, *J. Appl. Phys.*, 2001, **89**, 1393.
13. G. R. Bai, S. K. Streiffer, P. K. Baumann, O. Auciello, K. Ghosh, S. Stemmer, A. Munkholm, C. Thompson, R. A. Rao and C. B. Eom, *Appl. Phys. Lett.*, 2000, **76**, 3106.
14. V. Westphal, W. Gleeman and M. D. Glinchuk, *J. Appl. Phys.*, 1990, **68**, 2916.
15. A. E. Glazounov and A. K. Tagantsev, *Appl. Phys. Lett.*, 1998, **73**, 856.
16. D. Viehland, S. J. Yang, L. E. Cross and M. Wuttig, *J. Appl. Phys.*, 1990, **68**, 2916.
17. E. D. Specht, H.-M. Christen, D. P. Norton and L. A. Boatner, *Phys. Rev. Lett.*, 1998, **80**, 4317; H.-M. Christen, E. D. Specht, D. P. Norton, M. F. Chisholm and L. A. Boatner, *Appl. Phys. Lett.*, 1998, **72**, 2535.

Effects of Sr Substitution in Lead Zirconate Titanate Ceramics

H. Zheng, I. M. Reaney and W. E. Lee

Department of Engineering Materials, University of Sheffield, Sheffield S1 3JD, UK

N. Jones and H. Thomas

Morgan Electro Ceramics, Ruabon, Wrexham LL14 6HY, UK

ABSTRACT

$(Pb_{1-x}Sr_x)(Zr_{0.976-y}Ti_yNb_{0.024})O_3$ (PSZT) ceramics were investigated to understand the relationship between structural changes caused by Sr^{2+} substitution and dielectric, and piezoelectric properties. As Sr^{2+} was substituted for Pb^{2+}, the Zr:Ti ratio was modified so that compositions had an optimised piezoelectric coefficient (d_{33}). As Sr^{2+} content increased, optimised d_{33} also increased from 410 pC/N ($x = 0$) to 640 pC/N ($x = 0.12$), commensurate with a decrease in the paraelectric to ferroelectric phase transition temperature (T_C) from 350 to 175°C. [110] pseudocubic electron diffraction patterns revealed superlattice reflections at $^1/_2\{hkl\}$ positions associated with rotations of the octahedra in anti-phase. As Sr^{2+} content increased, sintering temperature had a pronounced effect on perovskite phase stability at the surface of PSZT. A second phase, formed in samples where $x = 0.12$ when sintered above 1170°C, was confirmed to be monoclinic ZrO_2, whose formation was accompanied by an increase in degree of tetragonality of the perovskite phase.

INTRODUCTION

PZT belongs to the ABO_3 perovskite structure, which consists of a corner-linked network of oxygen octahedra with Zr^{4+} and Ti^{4+} ions occupying the B-site within the octahedral cage and the Pb^{2+} ions situated in the interstices (A-sites) created by the linked octahedra. All compositions are cubic above the Curie temperature (T_C), but below, pure PZ transforms to an antiferroelectric (AFE) orthorhombic structure which persists with up to ~5%Ti^{4+} substituted for Zr^{4+}. Compositions between 5–47%Ti^{4+} are FE rhombohedral (R) whereas compositions with >47%Ti^{4+} are FE tetragonal (T). The boundary between T and R phases is temperature independent and known as the morphotropic phase boundary (MPB).[1]

Many compositional modifications to PZT have been developed. Donor doping, such as Nb^{5+} on the B-site, often leads to an enhanced ease of polarisation switching under ac field and these ceramics are referred to as "soft" PZT's,[2] while lower valent substituent, such as

Fe^{3+} on the B-site results in domain stabilisation and more difficult polarisation switching. These ceramics are termed "hard" PZT's and their piezoelectric properties do not change significantly with numbers of cycles.[3] With respect to undoped and "hard" PZT, "soft" PZTs have low coercive electric fields ($E_C < 2$ kV mm^{-1}), high remnant polarisation ($P_r > 0.3$ μC mm^{-2}) and high values of d_{33} ($d_{33} > 300$ pC/N). This soft behaviour is believed to arise from optimisation of extrinsic contributions to d_{33}, such as domain wall motion under the action of electric field or mechanical stress. There are three methods by which extrinsic contributions may be maximised in PZT:

(i)   donor doping with cations such as Nb^{5+} on the B-site. This gives rise to cation vacancies, limiting the oxygen vacancy mobility, thereby preventing pinning of domains by oxygen vacancy defect dipoles

(ii)   proximity to the MPB which decreases the free energy between R and T phases so that structural re-arrangement via domain wall movement is easier

(iii)   reducing the T_C by substituting less polarisable species such as Sr^{2+} onto the A-site. The total strain energy of the domain walls is diminished thereby reducing the activation energy for domain wall motion.

In this work, soft PZT ceramics based on Sr^{2+} substitution on the A-site were investigated. Their electrical properties were characterised and related to their phase assemblage and microstructure using XRD (X-ray diffraction), SEM (scanning electron microscopy), TEM (transmission electron microscopy) and EDS (energy-dispersive spectroscopy).

EXPERIMENTAL

Pellets of the solid solution $(Pb_{1-x}Sr_x)(Zr_{0.976-y}Ti_yNb_{0.024})O_3$ at various x and y values were prepared. Polycrystalline ceramic materials were obtained from reagent-grade raw materials: PbO, ZrO_2, TiO_2, Nb_2O_5, $SrCO_3$. The starting reagents weighed in appropriate ratios with addition of 2 wt.% excess PbO were mixed in distilled water and attrition-milled for 2 h. The powder obtained was calcined 4 h at 925°C, at a heating rate of 75 K h^{-1} and a cooling rate of 300 K h^{-1}. The calcined powder was attrition-milled again for 2 h with addition of 2 wt.% Carbowax PEG 10K as a binder. Dried and sieved powders were pressed uniaxially at 180 MN m^{-2} into 10 mm diameter pellets. The binder was burned off at 600°C for 3 h. Pellets were then embedded in 90 wt.%$PbZrO_3$ + 10 wt.%ZrO_2 powder and fired in closed alumina crucibles from 1140 to 1280°C for 4 h. The heating and cooling rates were 180 and 360 K h^{-1} respectively. All the samples made in this study had relative densities $\geq$95%.

An X-ray diffractometer with Cu K_α source, operated at 50 kV and 30 mA, was used for identification of phases and measurement of lattice parameters. A step size of 0.02°, a scan rate of 2°/min, and scan ranges of 20–60° (2θ) were adopted. Indices of all the XRD peaks were based on a pseudocubic cell. Unit cell constants for the tetragonal structure were calculated using (200) and (002) peaks.

A Camscan Series II SEM was used to examine grain morphologies in SEI (secondary electron imaging) mode. SEM samples were thermally etched 100°C below the sintering temperature for 0.5 h and then coated with gold before examination.

Samples for TEM were ground parallel to ~25 μm thickness. A 3 mm diameter Cu support ring with a 1 mm diameter hole was glued onto the ground ceramic. Samples were then ion thinned using a Gatan dual ion mill (Model 600), operating at 6 kV with a beam current of 0.3 mA per gun, and milling incidence angles of 10–15°. Specimens were coated with carbon before bright-field (BF) TEM examination using a Tecnai 20, operating at 200 kV, and a JEOL 3010, operating at 300 kV. EDS X-ray detectors and Link ISIS analysis systems, fitted to the TEMs, were used to analyse local chemical composition.

A LCR meter (Model 4284A) was used in conjunction with a computer-controlled temperature chamber to measure capacitance as a function of temperature. Measurements were taken at a frequency of 1 kHZ in the temperature range from 25 to 450°C on heating as well as cooling at a rate of 2 K min^{-1}. Capacitance was converted to dielectric permittivity, using the sample geometry and the permittivity of air.

Samples for d_{33} testing were sliced into ~0.8 mm thick discs and subsequently polished. Polished samples were electroded using silver paste and poled in insulation oil at 100°C for 3 min at 2 kV mm^{-1}. Measurements were then performed, 24 h after the poling, using a Piezometer System PM 2C at a frequency of 100 Hz.

RESULTS AND DISCUSSION

Sr^{2+} substitution for Pb^{2+} on the A-site in PZT

Previous work revealed that doping a MPB composition ($x = 0$) with Sr^{2+} on the A-site resulted in an increased degree of tetragonality.[4] Therefore, as the Sr^{2+} content was increased, the Zr:Ti ratio was raised to retain compositions in the vicinity of the MPB (for an optimised d_{33}). Four d_{33}-optimised compositions with Sr^{2+} doping at 0, 8, 12 and 16 mol.% were investigated.

Phase structure
Figure 1 shows XRD spectra from four d_{33}-optimsed compositions with Sr^{2+} (x mol.) substitution for Pb^{2+} where $x = 0, 0.08, 0.12$ and 0.16. It is evident that these four compositions are all tetragonal with the presence of either peak splitting or a shoulder on the {002} peak. No peak splitting or shoulder occurs on the {111}, confirming that the rhombohedral phase is absent. It is worth emphasising that these four optimised compositions possess similar tetragonal phase assemblage regardless of Sr^{2+} content.

Selected area electron diffraction
Figure 2 shows selected area electron diffraction patterns along the pseudocubic [110] zone axes of samples in Fig. 1. The most intense reflections can be indexed according to the fundamental perovskite structure with $a = $ ~0.40 nm, consistent with the presence of T/R phases in the PZT. Figure 2a ($x = 0$) only contains spots associated with the fundamental perovskite lattice and there is no evidence of superlattice reflections. However, Fig. 2b–d

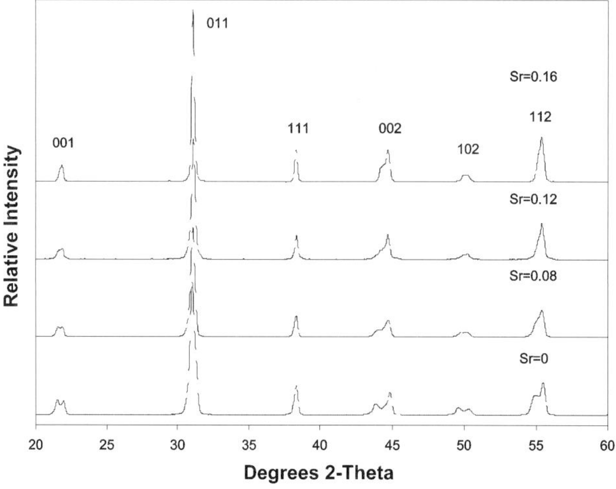

Fig. 1   XRD patterns from four d_{33}-optimised Sr-doped PZT ceramics.

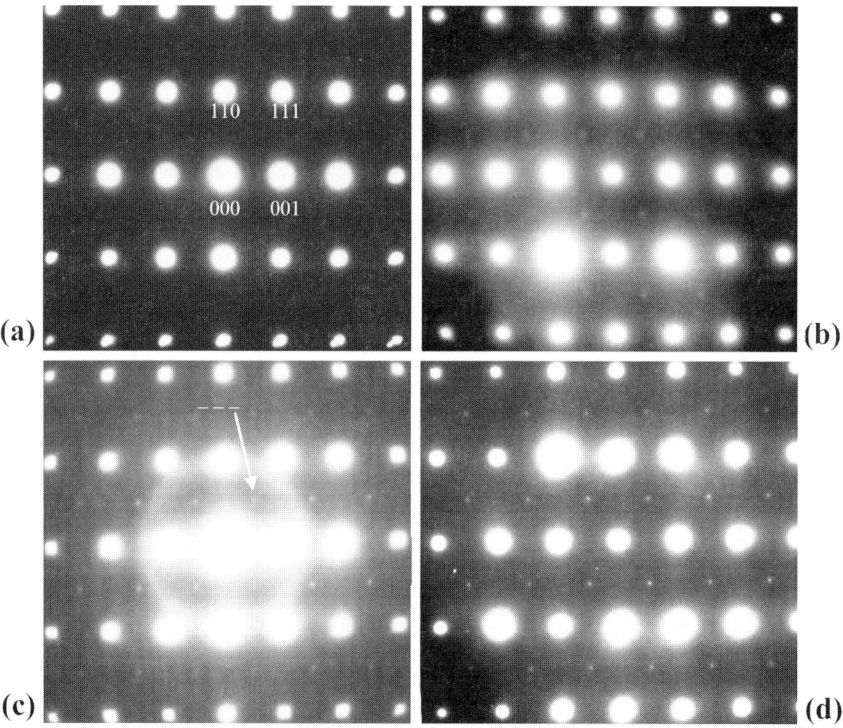

Fig. 2   Electron diffraction patterns along the pseudocubic [110] zone axes as a function of Sr^{2+} content cerrmics. (a) 0 mol.%Sr^{2+}; (b) 8 mol.%Sr^{2+}; (c) 12 mol.%Sr^{2+}; (d) 16 mol.%Sr^{2+}.

shows superlattice reflections occurring at the $\{h+^1/_2, k+^1/_2, l+^1/_2\}$ positions and their intensities become qualitatively stronger with increasing Sr^{2+} content.

In general there are three possible sources of superlattice reflections in stoichiometric perovskites:[5,6]

(i) ordering of cations. Cation species of differing valence states order themselves on $\{111\}$ planes perpendicular to a <111> direction of the pseudocubic perovskite cell. The resulting superlattice is doubled along all three principal axes. In diffraction, this is typified by the appearance of $^1/_2\{hkl\}$ reflections

(ii) antiparallel cation displacement. An AFE compound, such as $PbZrO_3$, is quadrupled along a <110> direction with respect to a simple pseudocubic cell, giving rise to $\pm^1/_2\{hk0\}$ reflections in diffraction patterns

(iii) tilting of the oxygen octahedra. Octahedral tilt transitions occur by the rotation of the octahedra around given axes of the pseudocubic cell. Glazer[6] described a notation based on rotations of octahedra for all possible tilt configurations. Anti-phase rotations are denoted as '–' and give rise to reflections which lie at the $^1/_2\{hkl\}$ position when $l = k \neq h$ in magnitude. In-phase rotations are ascribed the notation '+' and give rise to reflections which lie at the $^1/_2\{hk0\}$ positions when $h \neq k$ in magnitude.

For the case of ordering, a 1:1 ratio of dissimilar cations on the A- or B-site would give superlattice reflections of the type observed in Fig. 2 but there are insufficient Sr^{2+} ions (≤ 16 mol.%) to have a 1:1 cation ratio. Furthermore, ordering tends to occur when the charge and size differences between cations are large. The ionic size difference between Pb^{2+} and Sr^{2+} is small $(0.05\ \text{Å})^7$ and they are isovalent. Therefore, ordering is extremely unlikely at the source of the superlattice reflections. Compositions close to the MPB are all FE, and there is no evidence of AFE behaviour, anti-parallel cation displacement can also be dismissed. By elimination, it can be concluded that the superlattice reflections in Fig. 2 must arise from rotations of oxygen octahedra and based on the $^1/_2\{hkl\}$ type of superlattice reflections observed, the octahedral tilting occurs in anti-phase.

It is noticeable that superlattice reflections in Fig. 2b and c are weak and diffuse, and in Fig. 2d, they are weak but discrete. In general, weak intensities come from small scattering factor differences (small tilt angles) and diffuse reflections come from short-range ordered effects. Weak and diffuse implies that the amplitude of rotation is small and the correlation lengths over which they interact are also small. The weak, but discrete reflections in Fig. 2d imply a small amplitude of tilt but long-range order. It may be concluded, that the tilt transition temperatures in samples where $x = 0.08$ and 0.12 are effectively ambient in the microscope and thermal fluctuations result in diffuse reflections. For $x = 0.16$, the samples have undergone a phase transition above the ambient microscope temperature. Thermal fluctuations still exist but they never exceed the phase transition temperature and long-range order is maintained.

Anti-phase oxygen octahedral tilting

Megaw[5] and Glazer[6] concluded that reflections of the $^1/_2\{311\}$ type arose due to rotations in antiphase. In Fig. 2b–d, these $^1/_2\{311\}$ type reflections are present as weak and/or diffuse

Table 1   Tolerance factor data as a function of x values.

x	0	0.08	0.12	0.16
t	0.988	0.986	0.985	0.983

intensities However, Glazer's structure factor calculations suggest that the $1/2\{111\}$ type reflections, also present in Fig. 2b–d, are forbidden. Recent work by Reaney et al.[8] proposed that these reflections may appear in electron diffraction patterns by simple double diffraction

$$\tfrac{1}{2}\{311\} + \tfrac{1}{2}\{\bar{2}00\} \rightarrow \tfrac{1}{2}\{111\} \tag{1}$$

Reaney et al.[9] also suggested that antiphase rotations were likely to occur in compounds where the perovskite tolerance factor t is less than ~0.985. Table 1 lists the tolerance factor data at different x values calculated using data from Shannon and Prewitt.[7] Sr^{2+} (1.44 Å) is slightly smaller than Pb^{2+} (1.49 Å). In addition, the Zr:Ti ratio increases in MPB compositions with increasing Sr^{2+} to compensate for the induced increase in degree of tetragonality; Zr^{4+} is larger (0.72 Å) than Ti^{4+} (0.605 Å) and consequently, t decreases from 0.988 ($x = 0$) to 0.983 ($x = 0.16$). The t values ($t < 0.985$) given by Reaney et al.[10] for the presence of antiphase rotations of the octahedra were approximate and related to Ba and Sr based perovskites. Nevertheless, the values calculated in this study are in excellent agreement with those suggested by Reaney et al.[10]

Anti-phase oxygen octahedral tilting is commonly only observed in R PZT and has been widely reported.[11,12] However, there have been no studies relating the occurrence of antiphase oxygen octahedral tilting to the MPB or T phase PZT. Our studies suggest that antiphase oxygen octahedral tilting can also occur in T PZT.

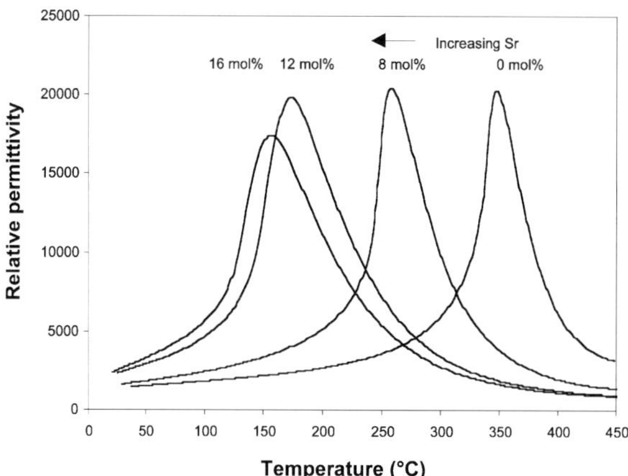

Fig. 3   Relative permittivities in four d_{33}-optimised Sr-doped PZT ceramics.

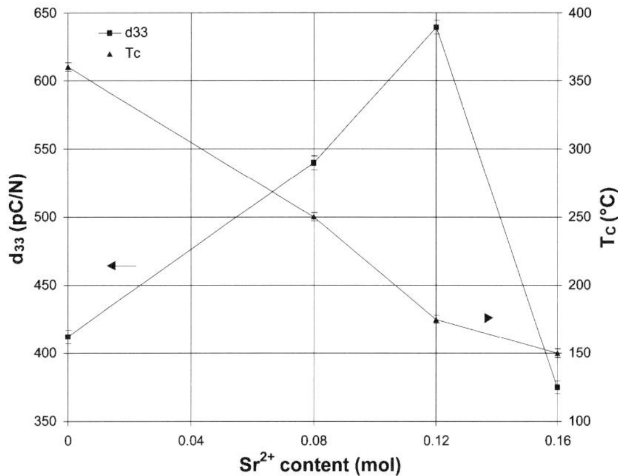

Fig. 4 d_{33} and T_C as a function of Sr^{2+} content in optimised Sr-doped PZT ceramics.

Electrical characterisation

Figure 3 shows plots of relative permittivity versus temperature for these MPB compositions. The T_C reduces from ~350 to ~150°C on doping with up to 16 mol.%Sr^{2+}. Broadening of the T_C maximum is also observed with increasing Sr^{2+} content, implying that compositional fluctuations become more evident at high Sr^{2+} content. It is assumed that the presence of microregions with local compositions (in the present context Pb^{2+}/Sr^{2+} and Zr^{4+}/Ti^{4+} ratios) varying from the average composition is responsible for the broadening of the T_C maximum.

Consistent with the decrease in T_C, the value of d_{33} increases with increasing Sr^{2+} concentration (Fig. 4). However, it reaches a maximum (~640 pC/N) at $Sr^{2+} = 0.12$, and then decreases for ceramics with $Sr^{2+} = 0.16$. The increase in d_{33} is commensurate with the decrease in T_C until $x = 0.12$ but the decrease for $x = 0.16$ does not fit this trend (Fig. 5). This result

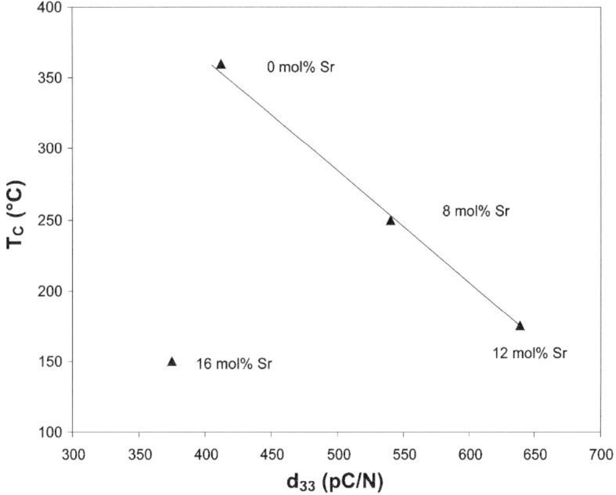

Fig. 5   TC versus d_{33} in optimised Sr-doped PZT ceramics.

disagrees with the premise that a lower T_C should reduce the activation energy for domain wall motion at room temperature, thereby softening the material.

As described previously, in ceramics with $x = 0.08$ and 0.12, the intensities associated with the superlattice reflections are diffuse and no long-range interaction of the amplitude of tilt occurs. The reflections observed in ceramics with $x = 0.16$ are discrete, implying long-range interaction. It is therefore proposed that for ceramics where $x = 0.16$, the octahedral tilt transition causes a long-range structural distortion which results in an increase in the strain energy of domain walls and therefore activation energy for domain wall motion. This would give rise to smaller extrinsic contributions to d_{33}. The short-range ordered tilting in ceramics where $x = 0.08$ and 0.12 will not cause further macroscopic distortion of the structure, neither increasing the strain energy of the domain walls nor their activation energies for motion.

Surface decomposition of Sr-doped PZT

In this study, the effect of sintering temperature on phase stability of $Pb_{0.88}Sr_{0.12}Zr_{0.538}Ti_{0.438}Nb_{0.024}O_3$ (PSZT12/44) ceramics was also examined.

Effect of sintering temperature on the stability of perovskite phase
XRD (Fig. 6) from the surface of sintered PSZT12/44 ceramics reveals that, although the tetragonal perovskite phase is always present after sintering for 4 h from 1070 to 1280°C, the degree of tetragonality increases with sintering temperature. After 4 h at 1070°C, asymmetry on the left side of the {002} peak occurs whereas clear splitting is evident after 4 h at >1120°C. Simultaneously, the intensity of the second phase peak at 27° (2θ) increases with increasing sintering temperature and further non-perovskite reflections emerge at T ≥1170°C. After 4 h at 1280°C, substantial second phase reflections are present. All non-perovskite reflections fit with monoclinic ZrO_2 with $a = 0.5151$ nm, $b = 0.5212$ nm and $c = 0.5317$ nm.[13]

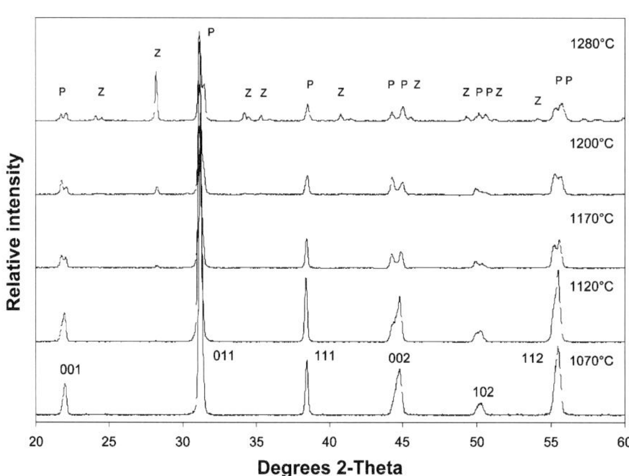

Fig. 6   XRD traces of PZT12/44 ceramics sintered 4 h at different temperatures: P is perovskite; Z is ZrO_2.

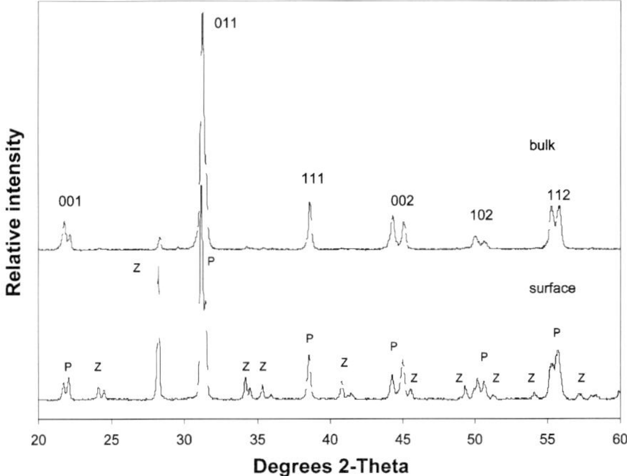

Fig. 7   XRD traces of bulk and surface PSZT12/44 ceramics sintered 4 h at 1280°C: P is perovskite, Z is ZrO_2.

The relative amounts of perovskite and second phase were quantified according to the following formula

$$P_p\% = \frac{I_{P_p}}{I_{P_s} + I_{P_p}} \times 100 \quad P_s\% = \frac{I_{Ps}}{I_{Ps} + I_{P_p}} \times 100$$

where $P_p\%$ and $P_s\%$ are the volume percentage of perovskite and second phases formed respectively, I_{Pp} is the relative intensity of the major X-ray peak for the perovskite phase, which has a d-spacing of 0.287 nm, and I_{Ps} is the relative intensity of the major X-ray peak for the second phase, which has a d-spacing of 0.316 nm.

Quantitative XRD indicated the sintered surface of PSZT12/44 ceramics contained 38 vol.% second phase. In addition, the *c:a* ratio as a function of sintering temperature of the perovskite phase was calculated. After 4 h at 1070°C the *c:a* ratio is 1. However, after 4 h at 1280°C, the *c:a* ratio of the sintered surface of PSZT is 1.02.

To determine whether decomposition of PSZT12/44 was limited to sintered surfaces or was present throughout the bulk, XRD was performed on samples from which the surface was ground to a depth of ~0.5 mm. Samples still exhibited a tetragonal perovskite phase but the intensity of the main ZrO_2 peak was greatly reduced, giving a calculated value for the percentage second phase of only 4% (Fig. 7).

A visual inspection of the pellets sintered 4 h at 1280°C in cross-section revealed a light contrast surface layer (~ 0.5 mm). To further investigate this surface phenomenon, SEM and TEM samples were prepared in cross-section.

Microstructure
Cross-sectional SEI SEM image of a thermally etched PSZT12/44 sample sintered 4 h at 1280°C revealed the change in morphology between surface and bulk. In the region close to

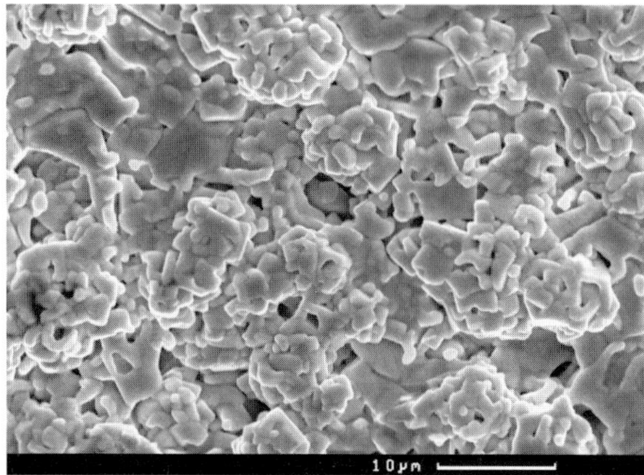

Fig. 8   SEI showing a region close to the PSZT12/44 surface, sintered for 4 h at 1280°C.

the surface, clusters of submicrometre (~0.8 mm) fine particles are dispersed within large particles, shown in detail in Fig. 8. SEM analysis also reveals high levels of porosity in this surface region consistent with PbO loss.

In the bulk of the sample, a conventional grain structure (3–4 mm in size), typical of PZT based ceramics,[14] is observed (Fig. 9). Compared with the surface (Fig. 8), levels of porosity are much lower in the bulk. Figure 10a is a BF-TEM image from a cross-section of a PSZT12/44 sample sintered for 4 h at 1280°C. Regions of perovskite adjacent to second phase particles were commonly observed. Typically, each second phase particle was surrounded by several perovskite grains, which were identified by electron diffraction pattern (Fig. 10b) and EDS (Fig. 10c).

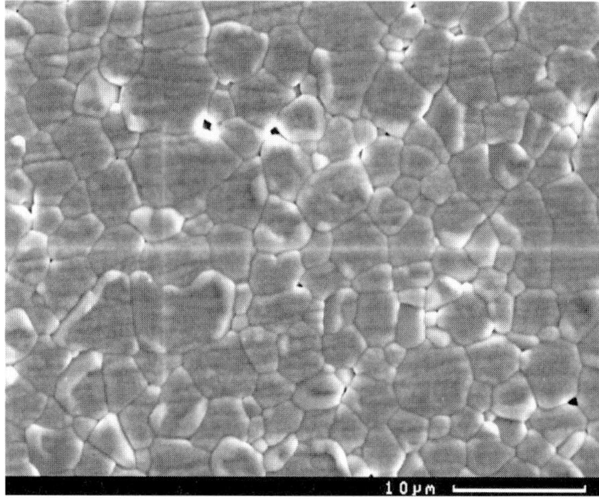

Fig. 9   SEI showing perovskite grains in the bulk of PSZT12/44 sintered for 4 h at 1280°C.

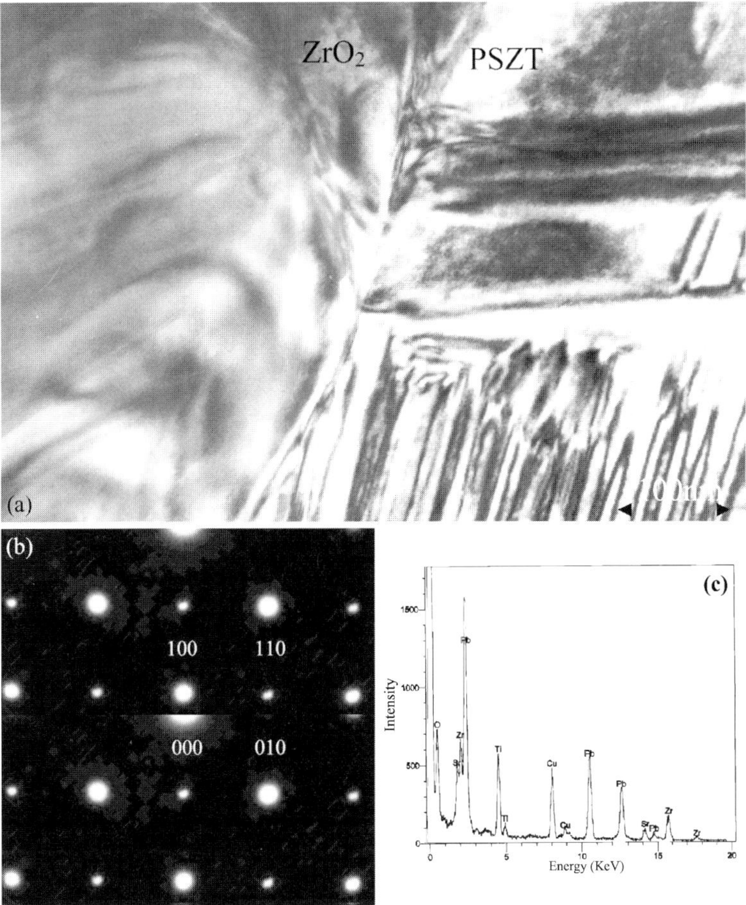

Fig. 10   (a) BF-TEM image showing the interface between a second phase ZrO_2 and perovskite matrix phase; (b) [001] zone axis electron diffraction pattern from a PZT perovskite grain; (c) EDS spectrum from a PZT perovskite grain in PSZT 12/44.

The second phase was determined to be pure ZrO_2 by electron diffraction (Fig. 11b) and EDS (Fig. 11c) and contained a ferroelastic domain structure, 100 nm in width, typical of monoclinic ZrO_2 (Ref.15 (Fig. 11a)). In contrast, the regions of PZT contained ferroelectric/ ferroelastic domains of about 30 nm in width (Fig. 10a). The lattice parameters of ZrO_2 phase calculated from its electron diffraction patterns are ~0.53 nm in good agreement with those determined by XRD.

3.2.3 Mechanism of decomposition

Two further facts help elucidate the decomposition reaction. First, PbO loss is known to occur from the surface of PZT at high temperatures.[16] This effect becomes increasingly significant as temperature increases and results in decomposition by ~1500°C. Second,

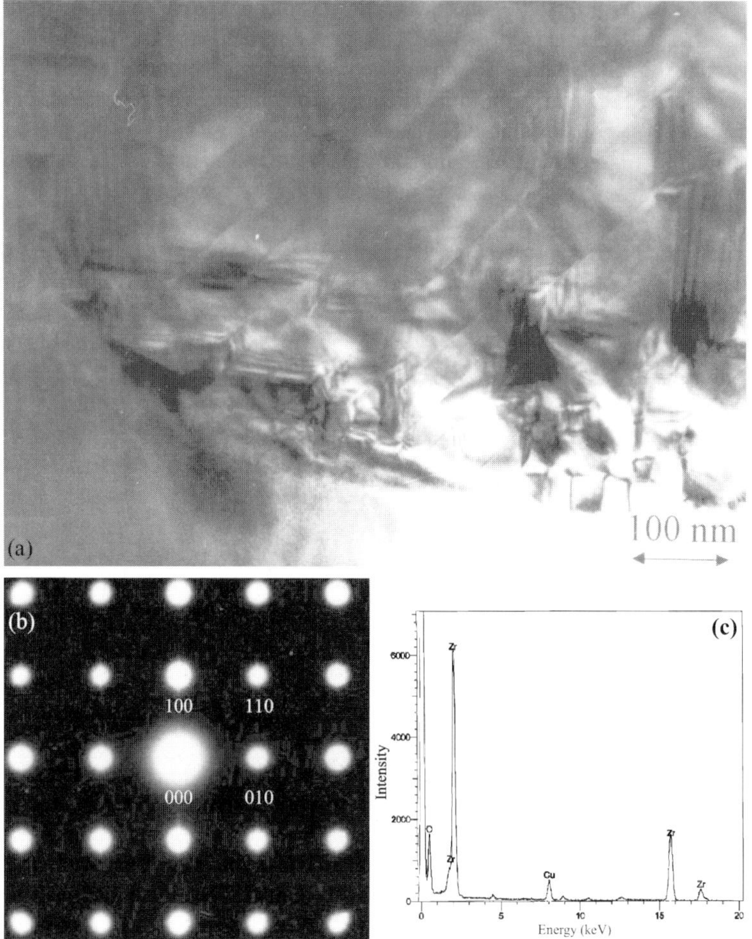

Fig. 11   (a) BF-TEM image showing typical ferroelectric domains in monoclinic ZrO_2; (b) [001] zone axis electron diffraction pattern; (c) EDS spectrum from ZrO_2 grain in PSZT12/44.

substitution of Sr^{2+} onto the A-site stabilises the T rather than R phase, and the Zr:Ti ratio must be increased to produce an MPB composition. PbO is therefore lost from the surface of PSZT not only due to its high volatility but also because Sr^{2+} favours a tetragonal perovskite structure, resulting in precipitation from solution of ZrO_2 particles simultaneously with an increase in tetragonality of the PZT.

CONCLUSIONS

Sr^{2+} doping in PZT greatly reduces the T_C, resulting in an increased d_{33}, which is optimised at ~640 pC/N for 12 mol.% Sr-doped PZT.

Sr^{2+} substitution on the A-site decreases the tolerance factor, resulting in the onset of the oxygen octahedral tilting even in T PZT.

Sintering temperature has a pronounced effect on the stability of perovskite phase in PSZT ceramics. A monoclinic ZrO_2 second phase forms as a result of PbO loss and Sr^{2+} substitution on the A-site promotes this decomposition.

REFERENCES

1. B. Jaffe, W. R. Cook and H. Jaffe, *Piezoelectric Ceramics*, Academic Press, 1971, 136.
2. Q. Tan and D. Viehland, *J. Am. Ceram. Soc.*, 1998, **81**(2), 328–336.
3. C. A. Randall, N. Kim, J. Kucera, W. Cao and T. R. Shrout, *J. Am. Ceram. Soc.*, 1998, **81**(3), 677–688.
4. H. Zheng, I. M. Reaney, W. E. Lee, N. Jones and H. Thomas, *J. Euro. Ceram. Soc.*, 2001, **21**(10/11), 1371–1375.
5. H. D. Megaw, *Crystals Structures: A Working Approach*, Saunders, 1973, 285–302.
6. A. M. Glazer, *Acta Crystallogr. B*, 1972, **28**, 3384–3392.
7. R. D. Shannon and C. T. Prewitt, *Acta Crystallogr B.*, 1970, **26**, 1046.
8. I. M. Reaney, A. Glazounov, F. Chu, A. Bell and N. Setter, *Brit. Ceram. Trans.*, 1997, **96**(6), 217–224.
9. I. M. Reaney, E. Colla and N. Setter, *Jpn J. Appl. Phys.*, 1994, **33**, 3984–3990.
10. E. L. Colla, I. M. Reaney and N. Setter, *J. Appl. Phys.*, 1993, **74**(5), 3414–3425.
11. X. Dai, Z. K. Xu, and D. Viehland, *J. Am. Ceram. Soc.*, 1995, **78**(10), 2815–2827.
12. D. Viehland, J. F. Li, X. H. Dai and Z. Xu, *J. Phys. Chem. Solids*, 1996, **57**(10), 1545–1554
13. R. H. French, S. J. Glass and F. S. Ohuchi, *Phys. Rev. B, Condensed Matter*, 1994, **49**(8), 5133–5142.
14. H. Zheng, I. M. Reaney, P. Y. Wang and W. E. Lee, *Ferroelectrics 2000*, eds. N. McN. Alford and E. Yeatmen, The Institute of Materials, 2000, 51–58.
15. J. M. Fernández, M. J. Melendo and A. D. Rodríguez, *J. Mater. Res.*, 1996, **11**(8), 1972–1978.
16. F. Fernández, C. Moure, M. Villegas, P. Durán, M. Kosec and G. Drazic, *J. Euro. Ceram. Soc.*, 1998, **18**(12), 1695–1705.

Modelling of 3–3 Piezocomposites for Hydrophones

H. Kara, A. Perry and C. R. Bowen

Materials Research Centre, Department of Engineering & Applied Science, University of Bath

ABSTRACT

Three-dimensional modelling of a 3–3 piezoelectric structure was carried out using ANSYS™ finite element modelling software. Hydrostatic figures of merit were calculated for structures with increasing amounts of interconnecting porosity present. In addition to air being the second phase, polymer fillers were added to the three-dimensional model in order to observe the effect of polymer Young's modulus on the piezoelectric properties of the bulk material. Results show that increasing the porosity has the effect of improving the hydrostatic piezoelectric properties for applications such as low frequency hydrophones.

THEORETICAL BACKGROUND

Traditionally, research into piezoelectric devices has concentrated on producing materials with high density. However, for certain applications it can be shown that ceramic piezoelectrics with low density, i.e. a high amount of open porosity, can have superior hydrostatic properties. One area of particular interest is that of low frequency hydrophones.[1]

At low frequencies (<100 kHz), the wavelength dimensions are greater than that of the hydrophone and the stress on the device due to the acoustic wave is effectively hydrostatic. For piezoelectric devices in active (driven) applications it is advantageous for the material to have a high d_{33} (strain per unit electric field in the direction of polarisation). Unfortunately, dense materials usually also have a high d_{31} (a contraction normal to the direction of polarisation). A figure of merit, which represents the hydrostatic strain per unit electric field, is used to assess the performance of a material in the active mode and is known as the hydrostatic strain constant, d_h.

$$d_h = d_{33} + 2d_{31} \ (\mathrm{m \ V^{-1}}) \tag{1}$$

For piezoelectric devices in passive applications (listening) the figure of merit, g_h, is used. This is defined as the electric field generated per unit hydrostatic pressure and is known as the 'hydrostatic voltage constant'.

$$g_h = \frac{d_h}{\varepsilon^T{}_{33}} \ (\mathrm{Vm^{-1} \ Pa^{-1}}) \tag{2}$$

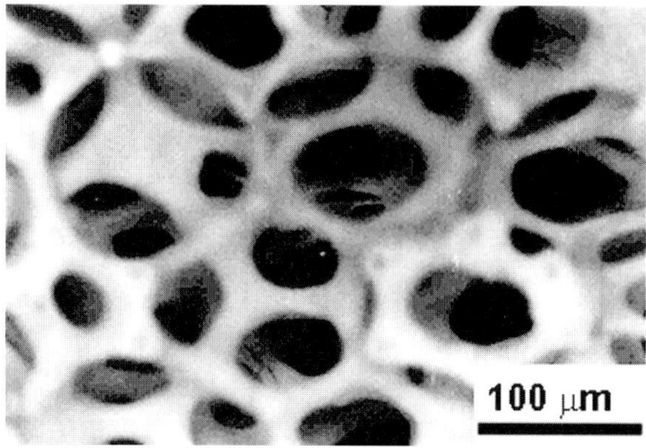

Fig. 1   An example of a 3–3 composite.

Both figures of merit, d_h and g_h, are related by the permittivity at constant stress, ε^T_{33}.

A third figure of merit is used in the case of the materials being used in both a passive and active mode. This is simply the product of the two figures of merit d_h and g_h, and is termed the hydrostatic figure of merit $(d_h \cdot g_h)$.

For dense materials the hydrostatic figure of merit, d_h, is low due to the fact that both d_{33} and d_{31} are large but of opposite sign. It is possible to increase the value of d_h by reducing the absolute value of d_{31}. This can be achieved by manufacturing a porous piezoelectric structure.[2,3] The mechanism by which d_{31} is reduced is explained in subsequent sections of this paper.

The research reported in this paper is concerned with 3–3 piezocomposites. This is a matrix of two materials (one is piezoelectric as an active phase, the other is polymer as a passive phase), that are completely interpenetrating, so that each phase forms a three-dimensional network around the other phase, as shown in Fig. 1.

In addition to increasing d_h, the inclusion of a second phase (air or polymer) reduces the overall permittivity, ε^T_{33}, of the device which increases the value of the piezoelectric voltage constant, g_h. Further gains in performance can be achieved by considering the density of the composite. A decrease in density of the device will result in lower acoustic impedance, leading to improved impedance matching between the acoustic medium (water or air) and the piezocomposite. Processing costs for open porosity piezocomposites can be considerably lower than those for dense materials or other types of composites, such as 1–3 structures. In addition, near net shape forming is possible and any post-sintering machining costs will be small.

A number of manufacturing methods can be used to produce a variety of porous structures at relatively low cost.[4] However, little work has optimised the performance of 3–3 piezocomposites with respect to pore content, pore morphology and properties of the passive phase. While models, including finite element models, have been developed to optimise other piezocomposites, such as 1–3 structures, no such model exists for 3–3 type

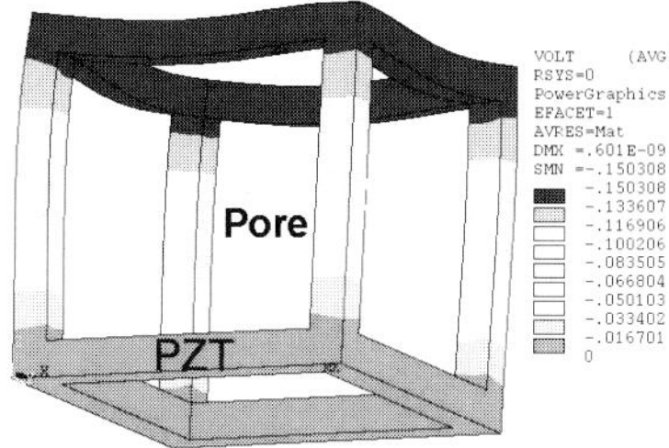

Fig. 2   Smaple unit cell used for FE modelling.

piezocomposites. Therefore, the current research is aimed at optimising the properties of 3–3 piezocomposites by using finite element modelling.

MODELLING

There are numerous physical properties that will have an effect on the final performance of a hydrophone device. The properties that have been examined are the volume fraction of porosity/polymer and the Young's modulus of the passive phase (polymer). The effect of changing these variables on the figures of merit d_h, g_h and $d_h.g_h$ will be discussed.

In order to model a porous structure, a unit cell was chosen to represent part of the structure which would be characteristic of the whole system. One such unit cell is shown in Fig. 2. In order to vary the porosity, the model was changed by increasing and decreasing the effective wall thickness of the unit cell. This enabled the model to be recalculated for any porosity from 0 to 100%, however, for practical reasons only porosities from 10 to 90% were calculated.

The model material chosen was PZT-5H, a commercially available soft PZT, and all piezoelectric data[5] relevant to this material was entered into a commercial finite element analysis (FEA) package. Electrodes were applied to the top and bottom of the unit cell and symmetry was applied to three of the six faces. This symmetry allowed the results from the unit cell to be calculated as if the cell was part of a larger array.

To calculate d_h, a potential difference was applied to the electrodes on the unit cell. The subsequent displacement and strain per unit field in the z direction (d_{33}) and the strain in the x or y (d_{31} or d_{32}) direction was measured (Fig. 3). These two values were used to calculate d_h, as in equation (1).

In order to calculate g_h, one electrode was set to 0 V, while the other electrode was free to attain an equilibrium potential (Fig. 4). A hydrostatic pressure was applied, acting to compress the structure, on the three surfaces of the unit cell. After the finite element model solution was calculated the voltage generated in the free electrode was measured allowing the calculation of g_h, the electric field generated per unit hydrostatic pressure. The product of the values d_h and g_h was used to calculate the hydrostatic figure of merit $d_h.g_h$.

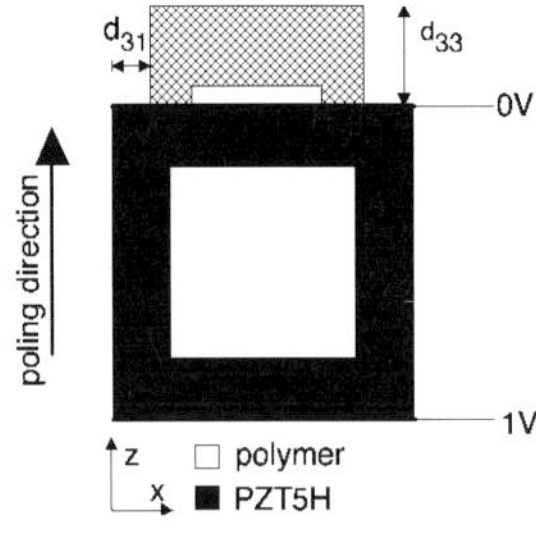

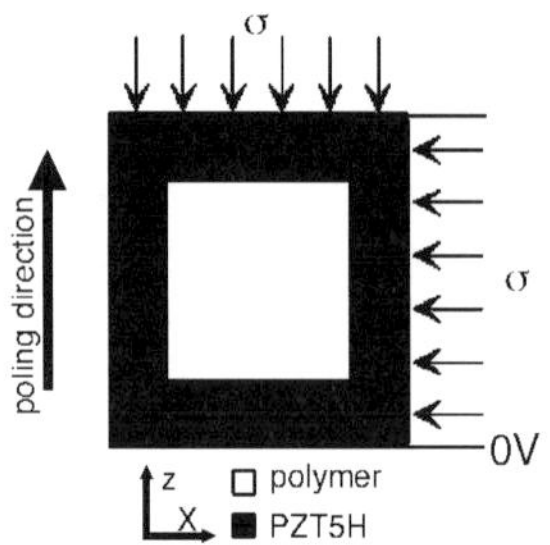

Fig. 3   Model used to calculate d_h.

Fig. 4   Model used to calculate g_h.

RESULTS

Effect of Young's modulus and polymer volume fraction on d_h

The following set of calculations was completed to assess the effect of volume fraction of polymer filled porosity on d_h. In addition, the Young's modulus of the passive polymeric second phase was varied from 2 to 10 GPa which include typical values of silicone polymers to hard set epoxies.

The results from this set of calculations can be seen in Fig. 5 which shows a maximum value of d_h at around 50% for the low modulus polymers and at around 35% for the high modulus polymers.

The behaviour in Fig. 5 can be explained in terms of 'active volume', as illustrated in Fig. 6. The term active volume is the volume of piezoelectric ceramic responsible for the strain in a particular direction. In the case of d_{33}, the active volume is the pillar of ceramic in the z direction. This volume decreases with increasing polymer volume content causing d_{33} to decrease at higher polymer volume content (Fig. 6a–c). Similar trends can be observed for d_{31} which rapidly reduces as the polymer volume fraction increases (Fig. 6d–f).

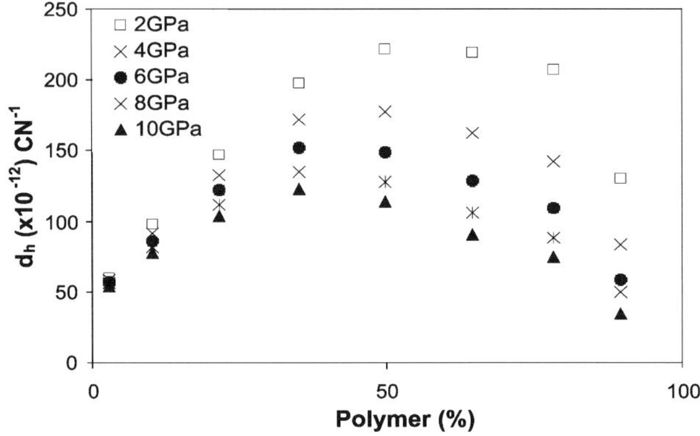

Fig. 5   Effect of polymer fraction d_h at various Young's moduli.

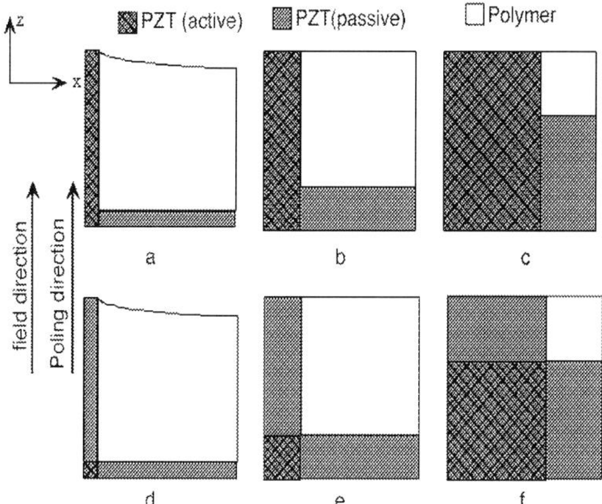

Fig. 6   Schematic of how d_{33} and d_{31} vary with polymer volume fraction.

However, the rate of reduction in d_{31} with increasing polymer content is faster than d_{33} as shown in Fig. 7. The d_{31} active volume is reduced due to the fact that only the volume of ceramic contained within the pillar in the z direction experiences a significant electric field. This is due to the low permittivity of the polymer compared with the ceramic ($\varepsilon_{\text{ceramic}}/\varepsilon_{\text{polymer}}$ ~400). The relatively slow reduction in d_{33} and a rapid reduction in d_{31} results in an increase in the figure of merit d_h to increase with increasing polymer volume fraction up to polymer volume fractions of 50%.

At higher polymer volume fractions there is insufficient piezoceramic for the polymer phase to strain with the ceramic, leaving a depressed region in the surface (Fig. 6a). As the value of d_{33} is calculated by averaging over the whole upper surface this depression ultimately reduces the value of d_{33} and subsequently d_h.

The value of d_h is also a function of the Young's modulus of the polymer phase and increases with decreasing Young's modulus. This is simply because a stiffer polymer will

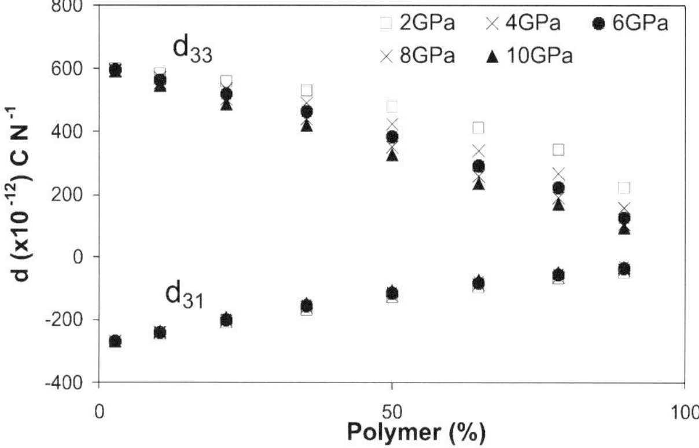

Fig. 7   Effect of polymer content on the d_{33} and d_{31} values.

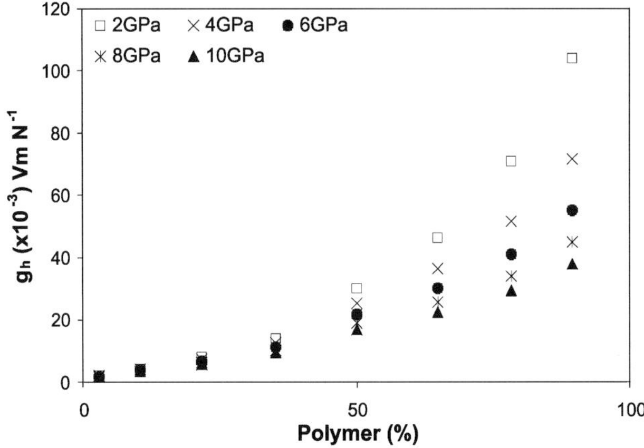

Fig. 8   Effect of polymer volume fraction and Young's modulus on g_h.

inhibit the movement of the active volume. From Fig. 5, it can be postulated that to maximise the performance of an active transducer a polymer volume fraction of around 50% with a low stiffness polymer should be used.

Effect of changing polymer volume fraction on g_h

A striking trend can be seen from Fig. 8 where g_h increases with increasing polymer volume fraction and reducing polymer stiffness. Figure 9 helps to explain this behaviour. The high stiffness of the ceramic relative to the polymer results in the majority of the applied stress in any direction being transferred into the ceramic. Thus, load per unit PZT area will increase with increasing polymer fraction. This results in high values of g_h (electric field per unit hydrostatic stress). As the polymer volume fraction increases the g_h value will rise. However, there is a limit where the ceramic part of the composite will fail or there will be critical stress where domain switching occurs. Therefore, a continuing increase in g_h, seen in Fig. 8, is actually prevented by practical considerations.

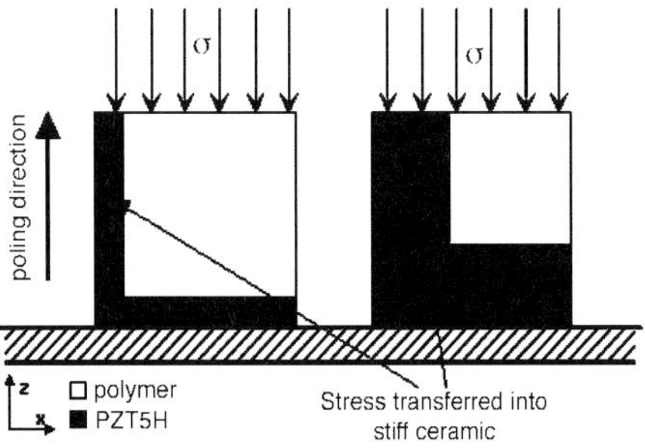

Fig. 9   Schematic of how g_h varies with polymer volume fraction.

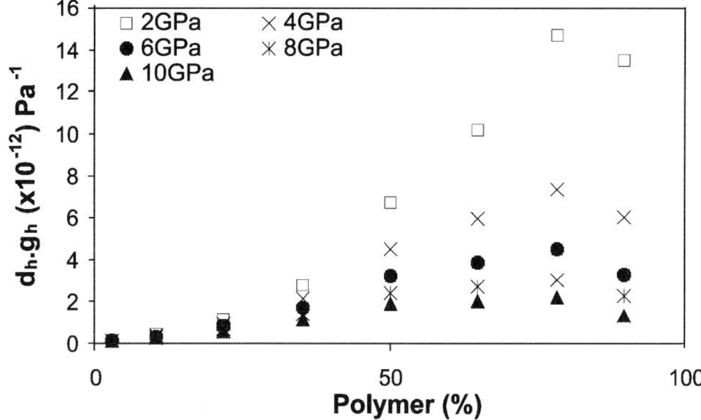

Fig. 10   Effect of polymer volume fraction and polymer Young's modulus on $d_h.g_h$.

The results for g_h show that in order to maximise the properties of a passive hydrophone device it is advantageous to have a high polymer volume content with a low stiffness polymer as a second phase. Our recent experimental results on PZT/polymer composites with varying polymer content showed a trend similar to FEA results.[6,7]

Effect of changing polymer volume fraction on $d_h.g_h$

Figure 5 and 8 have shown the effect of polymer volume fraction and Young's modulus on d_h and on g_h respectively. If a material is to be used in both passive and active mode, both of the figures of merit are of importance. These can be combined to give the hydrostatic figure of merit, $d_h.g_h$. This resulting curve is simply a combination of these two curves.

Figure 10 shows that for a passive/active device the optimum volume fraction porosity is around 82% for a low Young's modulus polymer filler and slightly lower at 75% for a high Young's modulus polymer filler. Optimum volume fraction is therefore a function of polymer stiffness.

CONCLUSIONS

There are a number of trends and conclusions that can be drawn from the finite element modelling results reported in this work.

- The figures of merit d_h, g_h and $d_h.g_h$ are enhanced as the Young's modulus of the second phase decreases for a fixed polymer volume fraction.
- The hydrostatic strain constant, d_h, reaches a broad maximum around 50% polymer volume fraction.
- The hydrostatic voltage constant, g_h, increases with decreasing ceramic volume fraction.

- The hydrostatic figure of merit, $d_h \cdot g_h$, which describes an active/passive transducer reaches a maximum at ~80% polymer volume fraction. Practical experiments are underway in order to quantify the accuracy of this model although the results are in good agreement with results reported in the literature.[8]

REFERENCES

1. P. Guillaussier and C.-A. Boucher, 'Porous lead zirconate titanate ceramics for hydrophones', *Ferroelectrics*, 1996, **187**, 121–128.
2. L. Montanaro, Y. Jorand, G. Fantozzi and A. Negro, 'Ceramic foams by powder processing', *J. Euro. Ceram. Soc.*, 1998, **18**, 1339–1350.
3. D. P. Skinner, R. E. Newnham and L. E. Cross, 'Flexible composite tranducers', *Mater. Res. Bull.*, 1978, **13**, 599–607.
4. C. Dias, D. K. Das-Gupta, Y. Hinton and R. J. Shuford, 'Polymer/ceramic composites for piezo-electric sensors', *Sensors Actuators A*, 1993, 37–38.
5. V. Yu Topolov and A. V. Turik, 'Non-monotonic concentration dependence of electromechanical properties of piezoactive 2–2 composites', *J. Phys. D: Appl. Phys.*, 2000, **33**, 1–13.
6. H. Kara, A. Perry, C. R. Bowen and R. Stevens, '3–3 piezocomposites: A comparison between the models and experimental results', *Ceram. Eng. Sci. Proc.*, **3/4**, to be published.
7. H. Kara, A. Perry, C. R. Bowen and R. Stevens, 'Interpenetrating PZT-polymer composites for hydrophones', submitted to *Ferroelectrics*.
8. K. Rittenmyer, T. Shrout, W. A. Schulze and R. E. Newnham, 'Piezoelectric 3–3 composites', *Ferroelectrics*, 1982, **41**, 189–195.

Lead-Doped Calcium Niobate–Tantalate Pyrochlores: Phase Structure and Dielectric Properties

J. C. Merry, A. C. Leach and R. Ubic

Department of Materials, Queen Mary, University of London, Mile End Road, London E1 4NS

ABSTRACT

The microwave properties of lead pyroniobates have been shown to be influenced by their crystal structures. The incorporation of PbO-rich layers as the concentration of Pb increases enables these materials to attain high Q_f but this is offset by poor ε_r and τ_f performance. Isovalent doping of Ta^{5+} ions on the niobium sites can reduce τ_f. In a similar study, calcium niobate–tantalate with zero τ_f was derived but both ε_r and Q_f were lower than for the lead pyroniobates. The present work aims to build on these previous findings by preparing pyrochlore ceramics, simultaneously doped on both A- and B-sites to produce ceramics with minimal τ_f values and enhanced quality factors for medium permittivity applications. A single-phase composition of $(Ca_{1.4}Pb_{0.6})(Nb_{0.5}Ta_{1.5})O_7$ was prepared by mixing pre-calcined batches of $Ca_2(Nb_{0.5}Ta_{1.5})O_7$ and $Pb_2(Nb_{0.5}Ta_{1.5})O_7$ in the appropriate ratio. Pellets were sintered to high densities and their dielectric properties tested. Preliminary data suggests that compositions with greater lead content and with a higher A- to B-cation ratio should be pursued.

INTRODUCTION

Oxide ceramics are critical components in microwave devices where they act as dielectric resonators, which may be used to determine and stabilise the frequency of a microwave oscillator or as a resonant element in a microwave filter. If the dielectric constant (ε_r) of the material is high enough, it can sustain a standing electromagnetic wave within its volume due to reflection at the air/dielectric interface, the frequency of which is dependent on the dimensions of the resonator, its permittivity and its environment. High quality factors (Q_f) allow for greater frequency tunability and through selective doping, ceramic components can be used to compensate for frequency drift due to their tunable temperature coefficients of resonant frequency (τ_f). By using ceramic dielectric resonators, the microwave device can be more compact and more temperature stable than those using resonator cavities.

This work aims to build on the findings of previous studies in applying appropriate doping to facilitate the production of an enhanced dielectric ceramic. Lead niobate pyrochlores

$(Pb_nNb_2O_{5+n}; 1.5 \leq n \leq 3.0)$ have been shown to display some useful dielectric properties.[1] High quality factors were observed for these materials but were coupled with poor ε_r and τ_f performance. The trend of decreasing ε_r and increasing Q_f was observed for $1.5 \leq n \leq 2.0$ and a minimum in τ_f of 814 ppm K^{-1} occurred for $n = 2.0$. This was related to the accommodation of PbO-rich layers in the crystal structure as the value of n increased. However, the value of τ_f was significantly reduced by the doping of tantalum ions onto the niobium sites.[2]

Similar work on calcium niobate–tantalate pyrochlores $(Ca_2(Nb_xTa_{2-x})O_7)$ has shown that $\tau_f \approx 0$ for $x = 0.36$ (Ref. 3). The end members of this system were shown to have τ_f values of similar magnitude but of opposite sign, and so such a balancing of τ_f values could have been expected. This phenomenon was closely related to the specific phase characteristics of these compositions. Although zero τ_f was achieved for these calcium niobate–tantalates, both ε_r and Q_f were lower than in the corresponding lead pyrochlores.

Therefore, by taking lead pyroniobate, and doping isovalently with both calcium and tantalum, it is anticipated that a balance in dielectric properties can be achieved. The aim of this work is to produce an enhanced pyrochlore ceramic with near zero τ_f and values of ε_r and Q_f that are commercially acceptable.

EXPERIMENTAL METHODS

Conventional mixed oxide powder processing techniques were used to prepare compositions in the CaO–PbO–Nb_2O_5–Ta_2O_5 system. Calcium niobate–tantalate was prepared by mixing stoichiometric amounts of $CaCO_3$ (99.9+%, Aldrich, Gillingham, UK), Nb_2O_5 and Ta_2O_5 (both 99.9%, H.C. Starck, Goslar, Germany), which were milled with 1 wt.% Dispex A40 (Allied Colloids, Bradford, UK) for 4 h and dried overnight at 80°C. The dried powder was then granulated with a mortar and pestle and sieved to under 500 mm. Calcination was achieved using a two-stage process. Firstly, the powder was heated to 1100°C for 1 h and weighed subsequently to ensure that all CO_2 evolved had been expelled from the batch. This was then re-milled for 4 h, dried and granulated before a second calcination at 1300–1450°C for 2 h. The process for lead-containing niobate–tantalates differed in that only a single calcination was necessary, at 750°C for 1 h, in a closed alumina crucible. PbO volatilisation was monitored by measuring the weight loss.

Reacted powders were then re-milled for a further 4–5 h, again with 1 wt.% Dispex A40, with 2 wt.% polyethylene glycol 1500 (Whyte Chemicals, London, UK) being added in aqueous solution 5–10 min before completion. Lead–calcium niobate–tantalates were prepared by mixing the appropriate quantities of the lead and calcium niobate–tantalates for 10–20 min. These slurries were then dried and granulated as above and subsequently pressed ($\approx$125 MPa) into cylindrical pellets, 10 mm in diameter and $\approx$3 mm thick. Sintering was conducted at temperatures between 1300 and 1500°C for 1–2 h. Pellets containing lead were sintered in closed alumina boats and were weighed before and after sintering to quantify the degree of PbO loss.

The phase assemblages of the calcined powders and sintered pellets were examined by X-ray diffraction (XRD). A Siemens D5000 Diffractometer (Siemens AG, Munich, Germany) was used, which employed Cu K_α radiation, $\lambda = 0.15406$ nm with a secondary monochromator. The scans were performed over a 2θ range of 10–60° with a step size of 0.02°, at a rate of 1 deg/min^{-1}.

Microstructural analysis was carried out in the scanning electron microscope (SEM) (JSM 6300, Jeol, Tokyo, Japan). Polished samples were thermally etched and coated with a thin ($\approx$100 nm) conducting layer of carbon prior to examination. Some pellets underwent thinning by conventional ceramographic techniques followed by ion milling (model 600, Gatan, Pleasanton, California, USA) to electron transparency for observation in the transmission electron microscope (TEM) (JEM 2010, Jeol, Japan).

Measurements of Q and τ_f were made at Filtronic Comtek (Wolverhampton, UK) and South Bank University (London, UK) on vector network analysers. Values of ε_r were calculated from the resonant frequency, obtained during the Q measurement.

RESULTS AND DISCUSSION

Calcium pyroniobate ($Ca_2Nb_2O_7$) pellets were sintered to near full density. The XRD results of these samples showed only $Ca_2Nb_2O_7$ peaks, with monoclinic symmetry. An examination of TEM micrographs of calcium pyroniobate and the selected area diffraction patterns of these grains (Fig. 1) clearly showed the presence of (100) twins, as reported previously.[4]

The calcined powder of calcium pyrotantalate ($Ca_2Ta_2O_7$) displayed the single-phase XRD pattern of the trigonal weberite structure reported by Grey et al.[5] Sintered pellets of this

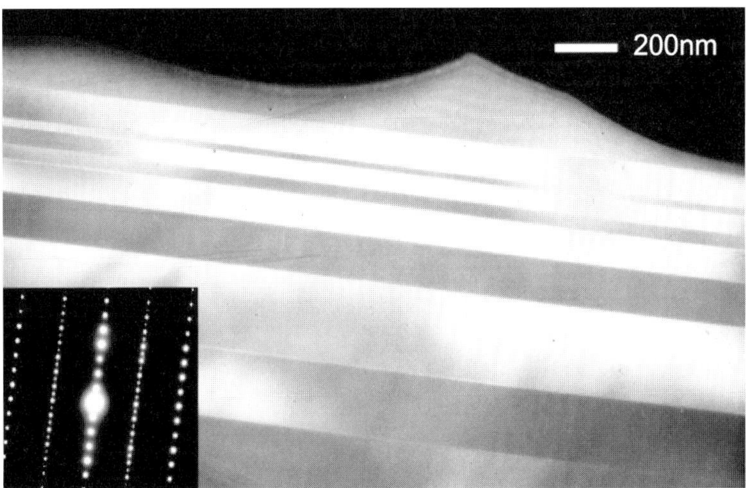

Fig. 1   Dynamical dark field image ($\bar{g} = 71\bar{1}$) of $Ca_2Nb_2O_7$ grain showing (100) twins and (inset) the corresponding [011] selected area diffraction pattern.

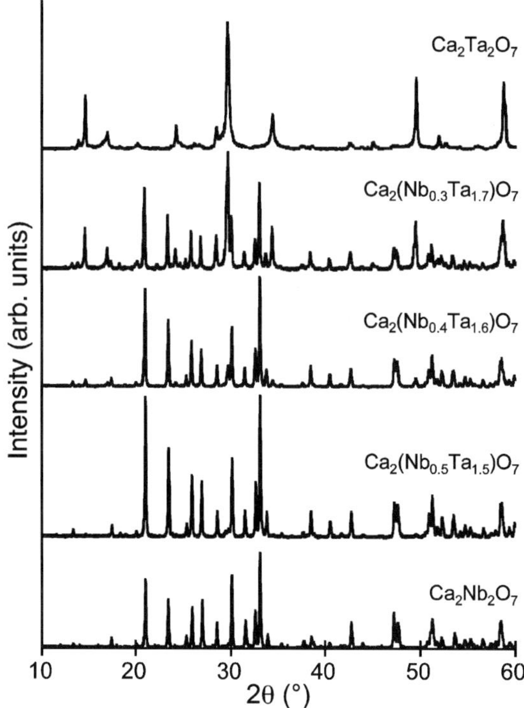

Fig. 2   XRD patterns for Ca$_2$(Nb$_x$Ta$_{2-x}$)O$_7$ composites x = 0, 0.3, 0.4, 0.5 and 2.

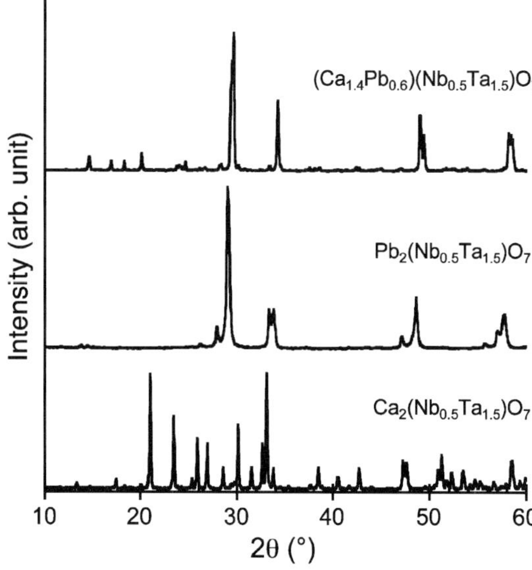

Fig. 3   XRD patterns sowing the phase characterisitcs of Ca$_2$(Nb$_{0.5}$Ta$_{1.5}$)O$_7$, Pb(Nb$_{0.5}$Ta$_{1.5}$)O$_7$ and (Ca$_{1.4}$Pb$_{0.6}$)(Nb$_{0.5}$Ta$_{1.5}$)O$_7$.

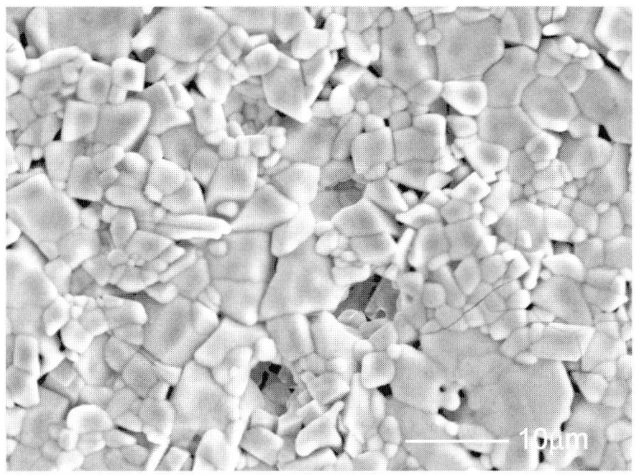

Fig. 4   Secondary electron micrograph of $(Ca_{1.6}Pb_{0.4})(Nb_{0.5}Ta_{1.5})O_7$ pellet sintered at 1450°C for 2 h.

composition displayed the characteristics of the monoclinic polytype, also reported by Grey et al. These pellets reached sintered densities of approximately 86% of the theoretical value using sintering temperatures up to 1500°C.

Mixed calcium niobate–tantalate compositions $(Ca_2(Nb_xTa_{2-x})O_7)$ were prepared for $x = 0$, 0.3, 0.4, 0.5 and 2. The XRD patterns of these compositions show predominantly single-phase calcium pyroniobate solid solution phases (Fig. 2). At $x = 0.3$ the XRD pattern shows a change in phase structure, with the calcium pyrotantalate-type solid solution becoming the dominant phase, but with a significant pyroniobate solid solution phase still apparent. These findings are consistent with the results of Cava et al. and Grey et al.'s studies. These compositions were sintered to densities in excess of 95% of the theoretical values.

The composition $(Ca_{1.4}Pb_{0.6})(Nb_{0.5}Ta_{1.5})O_7$ was prepared so that the beneficial level of tantalum doping[2] could be combined with levels of calcium that would promote near zero τ_f. The XRD patterns in Fig. 3 show that a single-phase composition was achieved. The phase structure of $(Ca_{1.4}Pb_{0.6})(Nb_{0.5}Ta_{1.5})O_7$ adopted the trigonal symmetry of the lead precursor rather than that of the calcium niobate–tantalate. Pellets of this composition were sintered to a maximum density of over 95% of the theoretical value at 1450°C; the microstructure of such a pellet is shown in Fig. 4.

The dielectric property data obtained for calcium pyroniobate and pyrotantalate served to confirm the findings of Cava et al.[3] However, where the previous study had measured these values at radio frequencies, the data in Table 1 was obtained at microwave frequencies. Pre-

Table 1   Dielectric properties data (corrected for porosity with Bötcher mixing rule).

Composition	ε_r	Q_f	τ_f (ppm K^{-1})
$Ca_2Nb_2O_7$	33.6	1300	-103
$Ca_2Ta_2O_7$	22.6	3300	235
$(Ca_{1.4}Pb_{0.6})(Nb_{0.5}Ta_{1.5})O_7$	39.2	3250	588

liminary testing of the lead–calcium niobate–tantalate composition showed a disappointingly high τ_f value. This was possibly due to the lack of the type of layered defect spacings in the crystal structure, the type that were so important to the dielectric properties of lead pyroniobates. To induce these kinds of defects the lead content and the A- to B-cation ratio must be increased.

CONCLUSIONS

A single-phase lead–calcium niobate–tantalate has been successfully produced with a trigonal lead pyroniobate-type structure. Work on calcium niobate–tantalate pyrochlores has confirmed what was previously observed in terms of phase structure and dielectric properties. In this study, mixed polytypes were observed in $Ca_2Nb_2O_7$ and the dielectric properties were measured at microwave frequencies. The preliminary dielectric data for $(Ca_{1.4}Pb_{0.6})(Nb_{0.5}Ta_{1.5})O_7$ signals an approach whereby lead content and, also, A- to B-cation ratio should be increased. In this way, it will be possible to produce an enhanced pyrochlore ceramic with useful ε_r and Q_f and near zero τ_f.

REFERENCES

1.   R. Ubic and I. M. Reaney, *Ceram. Trans.*, 2000, **106**, 263–275.
2.   R. Ubic and I. M. Reaney, *J. Eur. Ceram. Soc.*, in press.
3.   R. J. Cava, J. J. Krajewski and R. S. Roth, *Mater. Res. Bull.*, 1994, **33**, 527–532.
4.   J. F. Rowland, N. F. H. Bright and A. Jongejan, *Adv. X-ray Anal.*, 1960, **2**, 97–106.
5.   I. E. Grey, R. S. Roth, G. Mumme, L. A. Bendersky and D. Minor, *Mater. Res. Soc. Symp. Proc.*, 1999, **547**, 127–138.

Dopant Effects in Pyroelectric Ceramics in the PMN–PZT System

O. Molter, C. Shaw and R. W. Whatmore

Advanced Materials, School of Industrial and Manufacturing Science, Cranfield University, UK

ABSTRACT

This paper describes the properties of doped ceramics based on a solid solution of lead zirconate titanate (PZT) and lead magnesium niobate (PMN) that exhibit good pyroelectric performance. An optimal composition had been determined with uranium as a key-dopant for controlling electrical resistivity. Experimental studies have been conducted with the aim of replacing the uranium in this role. Ceramics were made using antimony as a donor dopant or chromium as an acceptor dopant. Measurements of resistivity, dielectric constant, dielectric loss and pyroelectric coefficient were carried out with two different electrode types, fired-on silver and evaporated gold–chromium. A variation of the pyroelectric properties with composition has been observed. Increasing the level of antimony resulted in slight increases in capacitance, dielectric constant and resistivity. However, loss was unchanged. Increasing the level of chromium resulted in significant decreases in resistivity and dielectric constant. However dielectric loss increases at high level of doping. The detailed properties of these donor and acceptor-doped ceramics are reported and compared with undoped material.

INTRODUCTION

There have been a number of studies on the pyroelectric properties of ferroelectric ceramics, such as lead zirconate titanate (PZT),[1–5] and their use in uncooled pyroelectric infra-red detecting and thermal imaging applications has been demonstrated.[6–9] A number of ternary systems including $Pb(Mg_{1/3}Nb_{2/3})O_3$–$PbZrO_3$–$PbTiO_3$ (Ref. 10) and Pb_2FeNbO_6–$PbZrO_3$–$PTiO_3$ (Ref. 11) have been studied and previous papers have dealt extensively with the effects of additives on the electrical properties.[12–14] It has been shown that the addition of uranium to Pb_2FeNbO_6–$PbZrO_3$–$PbTiO_3$ (Refs 15 and 16) and $Pb(Mg_{1/3}Nb_{2/3})O_3$–$PbZrO_3$–$PbTiO_3$ (Ref. 17) results in an electrical resistivity in the range 10^9 to 10^{11} Ω m^{-1}. Experimental studies on alternative donor and acceptor dopants have been conducted with the aim of replacing the uranium in this role.

EXPERIMENTAL PROCEDURES

Ternary lead zirconate–lead titanate–lead magnesium niobate (PZ–PT–PMN) ceramics with 0–3 mol.% donor (Sb_2O_3) or acceptor (Cr_2O_3) dopants were prepared by conventional ceramic technology.

Lead (II) oxide, titanium (IV) oxide, zirconium (IV) oxide, antimony (III) oxide, chromium (III) oxide and magnesium niobate[18] were used in the production of the ceramic. The magnesium niobate was prepared from basic magnesium carbonate $Mg(CO_3).Mg(OH)_2.5H_2O$ and niobium (V) oxide in an initial step using the route described by Butcher and Daglish.[19] Raw materials were wet-mixed by ball milling in water with Dispex® solution (0.1 wt.%). Following milling, the slurry was dried, sieved and then calcined at 800°C for 6 h in a covered alumina crucible in a muffle furnace. After sieving, the powder was remilled in water with Dispex® solution (0.1 wt.%). The material was dried and sieved. Binder solutions of Glascol® or 40/60 wt.% polyvinyl alcohol/polyethylene glycol were mixed with the powder using a combination of hand, ultrasonic bath and high shear (Silverson™) mixing. Pellets were pressed to 126 MPa using a 30 mm diameter, tapered, floating die arrangement.

Debinding and sintering were carried out in a muffle furnace. Pellets were placed uncovered on an alumina tray for the debinding process, which included 2 h dwell times at 250 and 600°C. For the sintering process, the pellets were covered with an alumina crucible to help prevent lead loss at the elevated sintering temperatures. Samples were sintered at 1250°C for 45 min. As-sintered pellets were examined using an optical microscope to ascertain the microstructure. To determine the electrical properties of material the pellets were electroded using two different techniques. One series was coated using silver paste and fired at 850°C for 5 min. The second series was electroded using evaporated gold–chromium.

Samples were poled using a Keithley high voltage source under an electric field of 3 kV mm^{-1} in hot mineral oil at 120°C for 10 min. The field was maintained until the oil temperature had cooled to 40°C. Pellets were washed for five minutes in acetone then in trichloroethylene for 5 min. They were placed overnight in an oven at 50°C with their electrodes shorted to remove any space charges that may have been introduced by the poling process. The dielectric properties of the electroded pellets were measured at 33 Hz using a GenRad 1689M RLC Digibridge. The pyroelectric current responses of the samples were measured using the Byer–Roundy method.[20] A computer controlled rig employing a thermoelectric heater/cooler was used to ramp the specimen temperature within the range 20–90°C under vacuum.

RESULTS

Density and microstructure

The average bulk density of the pellets after the sintering treatment was 7.7 g cm^{-3} which represents 96% of the theoretical X-ray density. The grain size of the undoped material, as judged from optical microscopy of the as-sintered surfaces, was about 10 mm. Additions of

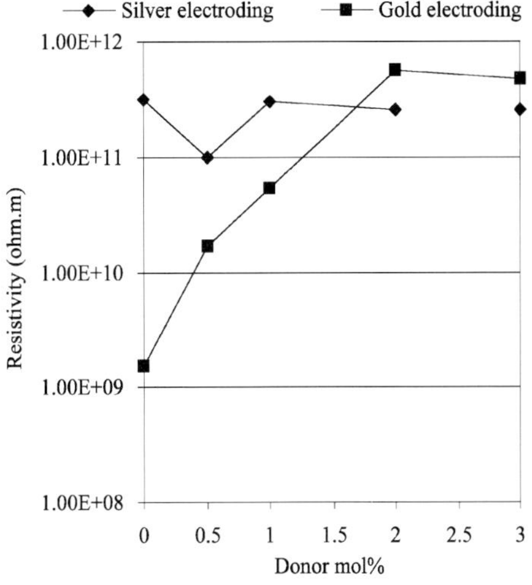

Fig. 1 Electrical resistivity of PZT–PMN ceramics + *x* mol.% donor additive.

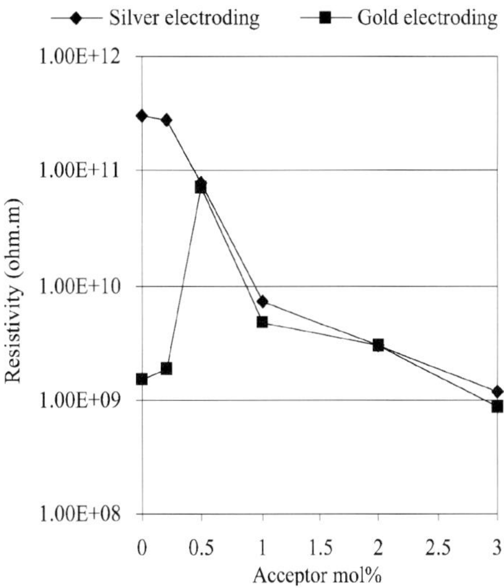

Fig. 2 Electrical resistivity of PZT–PMN ceramics + *x* mol.% acceptor additive.

up to 1 mol.% donor dopant increase the grain growth while higher levels of dopant lead to a reduction in grain growth. For the case of acceptor doping, there was a significant reduction in the grain growth with increasing dopant content.

Electrical properties

Resistivity

The resistivity is an important parameter for infra-red detection and thermal imaging applications since it will determine the electrical time constant of the front-end of the detector amplifier circuit. The level and the type of dopant affect the resistivity of the PZT–PMN materials, as shown Figs 1 and 2. For donor-doped ceramics, increasing dopant content tends to increase and stabilise resistivity in the region of 10^{11} Ω m^{-1}. This behaviour is particularly significant for gold-electroded materials. Resistivity of the silver painted ceramics does not change (10^{11} Ω m^{-1}). With increasing acceptor dopant level a significant decrease in resistivity from 10^{11} to 10^{9} Ω m^{-1} is observed with both electroding modes. Low resistivity has been observed for undoped and lightly-doped ceramics (below 0.5 mol.%) electroded with evaporated gold–chromium.

Dielectric constant

The dielectric constants of poled samples as a function of dopant additives are shown in Figs 3 and 4. For the silver-painted donor-doped samples, the dielectric constant has a minimum of 430 at 0.5 mol.% and then increases to 650 with increasing donor dopant content. The values are slightly lower for gold-electroded material but the general behaviour is the same

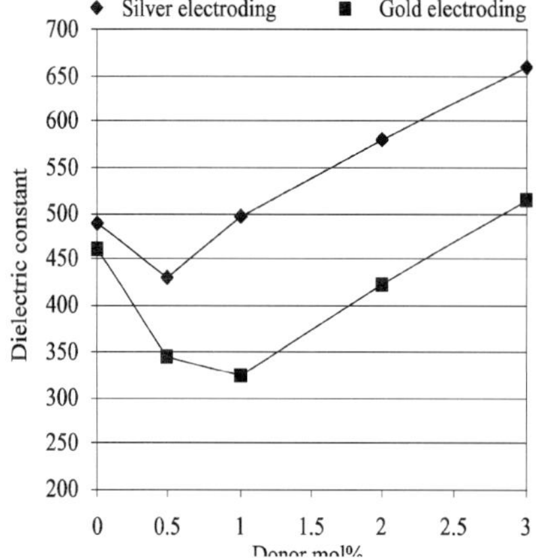

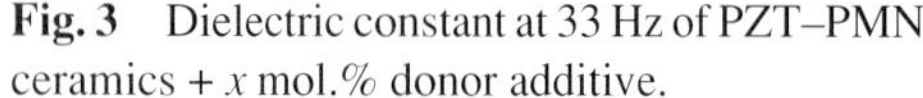

Fig. 3   Dielectric constant at 33 Hz of PZT–PMN ceramics + x mol.% donor additive.

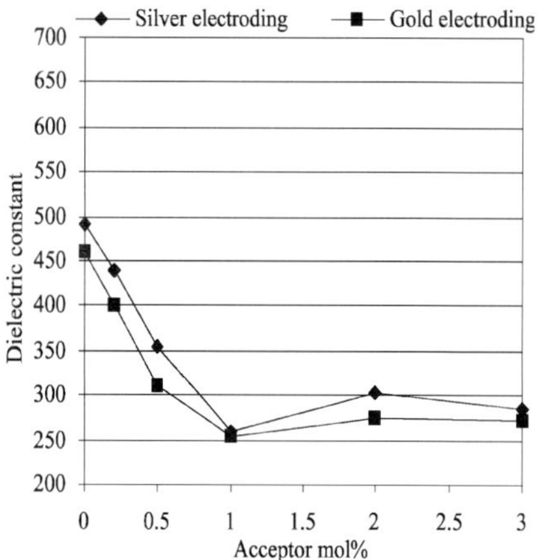

Fig. 4   Dielectric constant at 33 Hz of PZT–PMN ceramics + x mol.% acceptor additive.

with a minimum of 320 at 1 mol.% of dopant and increasing to 510 at 3 mol.%. For the acceptor-doped samples the dielectric constant decreases from 440 to 260 up to 1 mol.% then stabilises at ~300 over 1 mol.%. The same behaviour was observed with gold-electroded material. However the values of the dielectric constant were slightly lower.

Dielectric loss

The loss measured at 33 Hz for poled samples as a function of additives shows significant differences between the nature of the dopant, as shown in Figs 5 and 6.

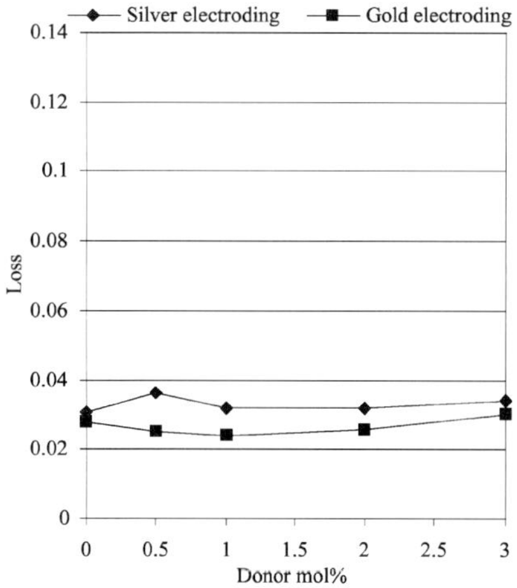

Fig. 5   Loss at 33 Hz of PZT–PMN ceramics + x mol.% donor additive.

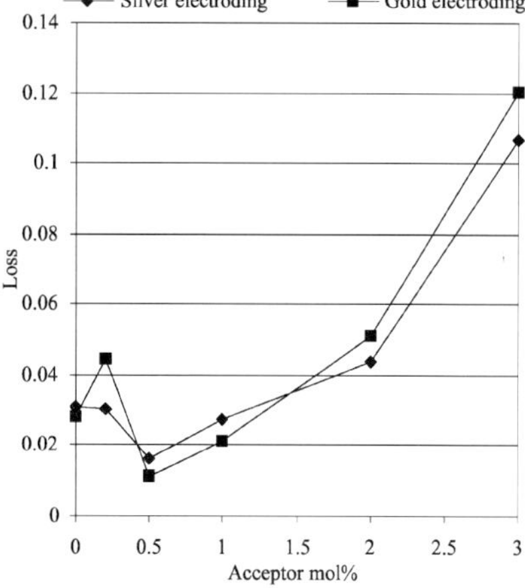

Fig. 6   Loss at 33 Hz of PZT–PMN ceramics + x mol.% acceptor additive.

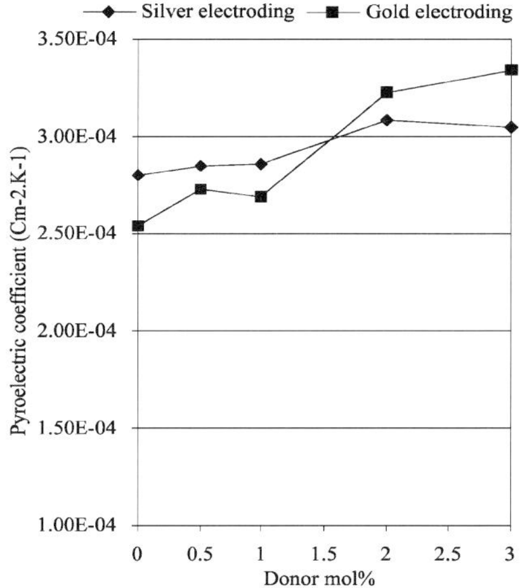

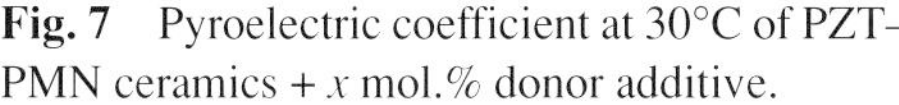

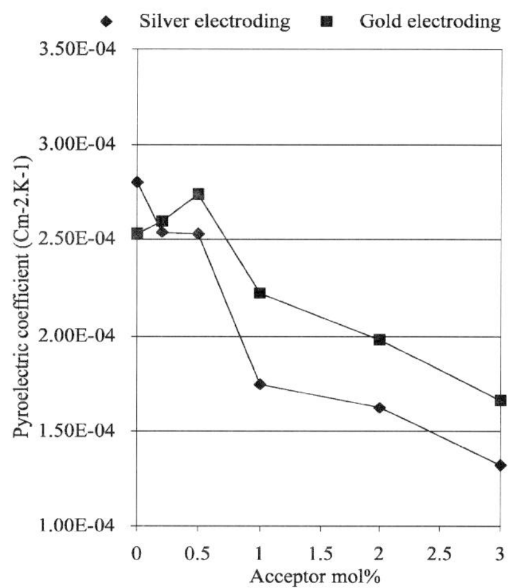

Fig. 7   Pyroelectric coefficient at 30°C of PZT–PMN ceramics + x mol.% donor additive.

Fig. 8   Pyroelectric coefficient at 30°C of PZT–PMN ceramics + x mol.% acceptor additive.

For the donor-doped ceramics, irrespective of the electrode type and doping level, the loss is approximately the same in the region 2.5–3%. A slightly lower loss for gold-electroded materials was observed. For the acceptor-doped ceramics, the level of doping significantly influences the loss of the material. Figure 6 shows a slight increase at low levels of doping (0.2 mol.%) with it being greater for gold-evaporated ceramics. A minimum is observed at 0.5 mol.%. Further increases in dopant lead to a rapid increase in loss to 12% at 3 mol.%.

Pyroelectric coefficient

The pyroelectric coefficient of each poled composition has been measured at 30°C and is shown for the donor- and acceptor-doped ceramics in Figs 7 and 8. The dopant effect is low with small amounts of additive (below 1 mol.% the values were similar for both of the additives). For donor-doped ceramics, increasing the dopant level leads to a slight increase in the pyroelectric coefficient. The effect is less significant for silver-painted than for gold-electroded pellets. For acceptor-doped ceramics, increasing the doping level apparently led to a general decrease of the pyroelectric coefficient common to the two modes of electroding.

DISCUSSION

Experimental studies have shown that varying the type and level of additive can vary the conductivity and the pyroelectric properties of PZT–PMN ceramics over several orders of magnitude.

Donor dopants also called 'softeners', such as penta-valent ions on the B-site or tri-valent ions on the A-site, donate electrons to the lattice. Antimony acts as a donor dopant in two possible ways with Sb^{3+} ions partially occupying A-sites in the lattice or via Sb^{5+} occupying B-sites.[21] However, the distribution between A- and B-sites has not been experimentally demonstrated. Donor doping leads to an increase of cation vacancies and a reduction of the concentration of oxygen vacancies giving a reduction in the concentration of domain-stabilising defect pairs. The number of free charge carriers in these ceramics also decreases, resulting from electron-hole compensation. The increase of resistivity of donor-doped ceramics is caused by electron-hole compensation with Sb^{3+} on the divalent A-site or Sb^{5+} on the tetra-valent B-site acting as electron donors. Lange and Kellet[22] have demonstrated that grain growth in a densifying powder compact is a result of inter-particle mass transport (i.e. coarsening) and is associated with grain boundary motion. Because of the complexity of mass-transport in multi-cation ceramics, the driving forces of grain boundary motion lead to abnormal grain growth and densification. The changes in dielectric constant of donor-doped PZT can be attributed to the changes in microstructure due to the increasing doping level. Roup[23] observed a similar increase in dielectric constant with donor-dopant levels in PZT. The observed increase in dielectric constant could be explained by the increase in the number of defects for high doping levels. Doping the PZT–PMN compositions with donor dopants also results in the increased mobility of the domain walls that contributes to increased dielectric constant. The reduction of oxygen vacancies can also give a small decrease in dielectric loss that has not been observed in this study.

Acceptor dopants also called 'hardeners', such as tri-valent ions on the B-site or monovalent ions on the A-site, accept electrons from the lattice. The effective valence of chromium is variable in ceramics and may even change with time and temperature. Vasileva[24] reported that Cr ions replace Zr or Ti ions in the lattice. They are hexa-valent at low concentrations but tri-valent at higher concentrations in ceramics. The observed changes in electrical properties may be due to the modification of chromium valency. Acceptors cause the existence of oxygen vacancies in the lattice. The significant difference between oxygen vacancies and cation vacancies in perovskite-type structures is the higher mobility of the former. Cations and cation vacancies tend to be separated by oxygen ions so there is a considerable energy barrier to be overcome before the ion and its vacancy can be interchanged. Oxygen ions, however, form a continuous lattice structure so oxygen vacancies have oxygen in neighbours with which they can easily exchange. The significant decrease of electrical resistivity may be due to the increase of the concentration of holes and thus an increase of the hole mediated conductivity. Takahashi[13] has shown a strong link between dielectric loss and resistivity for chromium-doped PZT ceramics. The increase of hole concentration may result in both an increase of dielectric loss and a decrease of electrical resistivity. The introduction of oxygen vacancies through acceptor doping also leads to a slight reduction in unit cell size, which tends to reinforce the effect referred to above. The shrinking and distortion of the cells by oxygen vacancies is thought to contribute to the lowering of the dielectric constant.

The significantly lower resistivities of undoped and lightly-doped ceramics of gold–chromium electroded materials relative to the fired-on silver electrodes have not been explained as yet. However, possible mechanisms may be associated with the diffusion of silver from the electrode into the ceramic.

CONCLUSIONS

Experimental studies on alternative donor and acceptor dopants have been conducted in the ternary $Pb(Mg_{1/3}Nb_{2/3})O_3–PbZrO_3–PbTiO_3$ system. Microstructure and the pyroelectric properties have been compared in the present paper for the two kinds of doping. The influence of donor or acceptor ions on the development of microstructure is very difficult to clarify and the link between microstructure and electrical properties of the different compositions has been suggested but not demonstrated.

Looking at the totality of the electrical results of the compositions studied, two levels of doping must be considered. Additions of less than 1 mol.% of donor or acceptor dopants seems to be very unstable and quite unpredictable. Resistivity depends significantly on the electroding mode, dielectric constant decreases and loss is generally stabilised between 0 and 1 mol.% dopant level. With dopant levels greater than 1 mol.% the electrical properties depend on the type of dopant. Increasing the level of donor-dopant over 1 mol.% resulted in slight increases in dielectric constant and resistivity however loss was unchanged. Increases in the level of acceptor-dopant over 1 mol.% resulted in a significant decrease in resistivity, an increase in dielectric loss and a little change in the dielectric constant.

ACKNOWLEDGEMENTS

The authors gratefully acknowledge EPSRC, IRISYS and the Royal Academy of Engineering for their financial support.

REFERENCES

1. B. Jaffe, W. R. Cook and H. Jaffe, *Piezoelectric Ceramics*, Academic Press, 1971.
2. R. Lane, D. Luff, K. R. Brown and H. J. Marshallsay, 'The variation of the pyroelectric properties with composition and phase structure for ceramics within the system lead zirconate titanate', *Trans. J. Brit. Ceram. Soc.*, 1973, **72**, 39–48.
3. D. Luff, R. Lane, K. R. Brown and H. J. Marshallsay, 'Ferroelectrics ceramics with high pyroelectric properties', *Trans. J. Brit. Ceram. Soc.*, 1974, **73,** 251–264.
4. Y. Iida, T. Ogawa and M. Toyoda, 'Pyroelectric characteristics of lead titanate zirconate family ceramics', *Electrical Eng. Jpn*, 1977, **97**(6), 1–7.
5. J. J. Dih and R. M. Fulrath, 'Electrical conductivity in lead zirconate–titanate ceramics', *J. Am. Ceram. Soc.*, 1978, **61**(9/10), 448–451.
6. J. M. Herbert, *Ferroelectric Transducers and Sensors*, Gordon and Breach, 1982.
7. R. W. Whatmore and F. W. Ainger, 'Pyroelectric ceramic materials for uncooled I.R. detectors', *Advanced Infra-red Sensor Technology*, ed. Jean Besson, International Society for Optical Engineering, 1983.
8. A. J. Moulson and J. M. Herbert, *Electroceramics*, Chapman & Hall, 1990.
9. R. W. Whatmore, 'Pyroelectric ceramics and devices for thermal infra-red detection and imaging', *Ferroelectrics*, 1991, **118**, 241–259.

10. H. Ouchi, K. Nagano and S. Hayakawa, 'Piezoelectric properties of $Pb(Mg_{1/3}Nb_{2/3})O_3$–$PbTiO_3$–$PbZrO_3$ solid solution ceramics', *J. Am. Ceram. Soc.*, 1965, **48**(12), 630–635.

11. R. W. Whatmore and A. J. Bell, 'Pyroelectric ceramics in the lead zirconate–lead titanate–lead iron niobate system', *Ferroelectrics*, 1981, **35**, 155–160.

12. H. Ouchi, M. Nishida and S. Hayakawa, 'Piezoelectric properties of $Pb(Mg_{1/3}Nb_{2/3})O_3$–$PbTiO_3$–$PbZrO_3$ ceramics modified with certain additives, *J. Am. Ceram. Soc.*, 1966, **49**(11), 577–582.

13. M. Takahashi, 'Space charge effect in lead zirconate titanate ceramics caused by the addition of impurities', *Jpn J. Appl. Phys.*, 1970, **9**(10), 1236–1246.

14. M. Takahashi, 'Electrical resistivity of lead zirconate titanate ceramics containing impurities', *Jpn J. Appl. Phys.*, 1971, **10**(5), 643–651.

15. A. J. Bell and R. W. Whatmore, 'Electrical conductivity in uranium doped modified lead zirconate pyroelectric ceramics', *Ferroelectrics*, 1981, **37**, 543–546.

16. R. W. Whatmore, 'High performance, conducting pyroelectric ceramics', *Ferroelectrics*, 1983, **39**, 201–210.

17. R. W. Whatmore, UK Patent 9904027.

18. S. L. Swartz and T. R. Shrout, 'Fabrication of perovskite lead magnesium niobate', *Mater. Res. Bull.*, 1982, **17**, 1245–1250.

19. S. J. Butcher and M. Daglish, 'The use of magnesium carbonate hydroxide pentahydrate in the production of perovskite lead magnesium niobate', *Proc. Third Euro-Ceramics*, eds. P. Duran and J. F. Fernandez, Vol. 2, ECERS, 1993, 121–126.

20. R. L. Byer and C. B. Roundy, *Ferroelectrics*, 1972, 3; *IEEE Trans. Sonics Ultrason.*, 1972, **SU-19**, 333.

21. G. H. Dai et al., 'A study of Pb vacancies and Pb–O vacancy pairs in doped $Pb_{0.85}Sr_{0.15}$ $(Zr_{0.55}Ti_{0.45})O_3$ ceramics by positron annihilation', *J. Mater. Sci.: Mater. Electron.*, 1991, **2**, 164, 170.

22. F. F. Lange and B. J. Kellet, 'Thermodynamics of densification : ii, grain growth in porous compacts and relation to densification', *J. Am. Ceram. Soc.*, 1989, **72**(5), 735–741.

23. R. R. Roup, H. R. Laird and H. U. Taylor, 'Composition céramique', Brevet d'invention, no.1,308,231, Bulletin officiel de la propriété industrielle, no.44, 1962.

24. T. K. Vasileva, M. M. Nadoliiski and S. D. Toshev, 'Study of PZT ceramics doped with Cr_2O_3, Physica Status Solidi (A)', *Appl. Res.*, **86**(2), K109–K111.

Aqueous Tape Casting of PZT Ceramics

A. Navarro, J. R. Alcock and R. W. Whatmore

School of Industrial and Manufacturing Science, Cranfield University, Cranfield, UK

ABSTRACT

PZT ceramics have been tape cast into thin ceramic sheets (200 mm) using water as the carrier fluid. The ceramic particles were stabilised using a commercial polyacrylic acid (D3021). The optimum dispersant level was obtained by viscosity measurements. Polyvinyl alcohol (PVA) was used as the binder and polypropylene glycol (PPG 400) as the plasticiser. Tensile testing of the green tapes found that 35 vol.% PVA+PPG400 is sufficient for handleable tapes. The amount of water in the slip was varied between 75.3 to 69.7 vol.% to find the optimal water content needed to permit faster drying of the tapes at room temperture. Lower water content produced higher viscosity slips, which facilitated slower sedimentation of particles during drying and more reproducible dried tape thicknesses.

INTRODUCTION

Tape casting[1,2] is mainly used for manufacturing thin, flat ceramic sheets which can be cut or punched to various shapes or sizes and stacked or laminated as multilayered structures. Applications of the tape casting approach include tapes being used to make electronic devices,[3] such as multilayer capacitors, and to produce insulating substrates. Newer applications include piezoelectric actuators, tranducers, solid oxide fuel cells and pyroelectric infra-red detectors based on lead magnesium niobate (PMN) doped PZT ceramics. Traditionally, tape casting has followed a non-aqueous route since organic solvents have low latent heat of evaporation and low surface tension. However there is an expected transition towards water-based systems because of environmental, health and safety, and economic reasons. Hotza and Greil[4] outlined a rewiew on aqueous tape casting. A few examples of non-aqueous ceramic systems include Si_3N_4, PZT and $BaTiO_3$ reported by Gutierrez and Moreno,[5] Galsassi et al.[6] and Mackinon and Blum,[7] respectively. Recently, aqueous tape casting has been developed for some ceramic systems, such as alumina[8,9] AIN[10] and PLZT.[11] Possible disadvantages of the aqueous route include increased stress sensitivity of tapes, increased flocculation of slips, slower drying leading to increased sedimentation and segregation of particles, poor wetting of slips due to the high surface tension of water and in the case of $BaTiO_3$ (Ref. 12), reactions with water.

The role of slip additives in tape casting has been investigated by Moreno.[13,14] By introducing polymer additives, a stable slip with the appropriate rheological properties can be achieved to produce homogenous tapes with high green density. A wide range of water-based binders exist which can be used for tape casting, ranging from acrylic latex emulsions[15] to cellulose ether deriveratives.[16] One common binder is poly(vinyl alcohol) which has been used in die-pressing and spray drying of ceramic powders.

EXPERIMENTAL PROCEDURES

Materials

PMN doped PZT powders of composition composition $Pb[\{(Mg_{1/3}Nb_{2/3})_{0.075}(Zr_{0.925}Ti_{0.075})_{0.925}\}_{0.99} Mn_{0.01}]O_3$ were produced for tape casting. The powders were made using the mixed oxide route. Powders with average particle size of 0.5 mm and specific surface area of $1.209 \ m^2 \ g^{-1}$ were produced for tape casting experiments. The dispersant was an ammonium salt of a polyacrylic acid, Duramax™ D3021, $M_w = 2500$ (Rohm and Haas, France), which can dissociate in water to produce a negatively charged polyion (-RCOO⁻) and NH_4^+ counterions, which remain in solution. Poly (vinyl alcohol) (PVA, 87–89% hydrolysis, $M_W = 105,000$) (Aldrich, UK) was used as the binder and poly (propylene glycol) (PPG) 400 (Aldrich, UK) as the plasticiser. A wetting agent (Surfynol SE-F, Air Products and Chemicals, Holland) was also added to the slips to aid wetting of the slurries.

Zeta potential measurements

Dilute PZT suspensions (1 vol.%) were prepared without a dispersant at different pH values. Acetic acid (1 M) and ammonium hydroxide (1 M) were used as the titrants to adjust the pH accordingly. The suspension was placed in 250 mL silica milling jars with cyclindrical zirconia media and ball milled for 18 h to break up any agglomerates present prior to the measurements. Zeta potential measurements were performed using a Malvern Zetasizer 3000. This method uses the principle of electrophoresis, which is the measurement of the movement of colloidal particles in the presence of an electric field. This is related to the surface charge on the particle and so the zeta potential. The drawback of this method is that particles must undergo Brownian motion in the cell during the measurement. Therefore samples were also ultrasonicated for 5 min to disperse the particles uniformly before the analysis.

Viscosity measurements

Viscosity measurements were conducted using a Couette head rheometer (Bohlin CS, Cirencester, UK). A pre-shear of $150 \ s^{-1}$ followed by rest was conducted for each suspension to erase any shear history inherent in the samples. The suspensions were also allowed to reach thermal equilibrium before each measurement. 20 and 30 vol.% PZT suspensions were ball milled for at least 18 h. The suspensions contained various concentrations of Duramax D3201 dispersant (Rohm and Haas, France). The viscosity was measured at a single shear rate of $100 \ s^{-1}$. This procedure was repeated three times at regular 10 min intervals to obtain an average value for the viscosity.

Tape casting

Tape casting slips were prepared by ball milling PZT, deionised water and 1.0 wt.% of D3201 dispersant for approximately 8 h. Surfynol SE-F wetting agent was then added to the slip, which was ball milled for 1 h. PVA was pre-dissolved in a suitable amount of water according to the method described by Bassner and Klingenberg.[17] This was achieved by heating the PVA solution to 95°C for at least 2 h. PPG 400 was then added to the slip and ball milled for a further 18 h to ensure a well-dispersed slurry. Tape casting was performed using a laboratory-scale, tape caster (TTC-1000, Mistler Inc., USA) with a stationary casting head. The carrier film used for tape casting was untreated BOPP film (Western Wallis, USA) with a surface energy of 31 dynes/cm^3.

Tensile testing

The tensile testing of green tapes with 39, 37, 36 and 35 vol.% PVA+PPG was accomplished using an Instron universal testing machine (Model 6025, Instron, UK). The binder and plasticiser ratio (%) was kept constant at a 50:50. A 500 g load cell was used. Studies by Tanaka et al.[18] showed that moisture can act as a plasticiser for PVA binders by affecting the glass transition temperature (T_g). In addition, the green density and green strength of the tapes is also affected. To prevent additional plasticisation, the green tapes were stored in a sealed polyethylene bag at room temperature (20°C ± 1) and a measured humidity of 35 ± 5% (humidity meter) preceding tensile testing. Dog bone shape tensile specimens were cut from portions of green tapes using a razor (7 mm width and 90 mm gauge length). Firm 12 mm^2 cardboard squares with double-sided sticky tape were attached to the top and bottom grip sections of the bar specimen to prevent tearing during testing. A crosshead speed of 5 mm min^{-1} was used. At least 10 specimens were tested for each measurement to obtain an average value for the tensile properties, but samples that tore during testing were excluded from the results.

Thickness measurements of green tapes

Four slips were produced at 75.3, 73.3, 72.1 and 69.7 vol.% water respectively, using 1.0 wt.% D3201 and 35 vol.% PVA+PPG. The tapes were cast at a constant gap height of 250 mm and a casting rate of 0.75 cm s^{-1}. To determine the thickness of dried tapes, at least 10 (30 x 30 mm^2) tapes were cut out from various sections of the tape and the thickness was measured using a flat-faced micrometer.

RESULTS AND DISCUSSION

Colloidal processing of PZT suspensions

Figure 1 shows the zeta potential of 1 vol.% PZT suspension as a function of pH. The isoelectric point corresponds to a zeta potential of zero, therefore IEP ≈ 6.5. Below pH 6.5, the particles

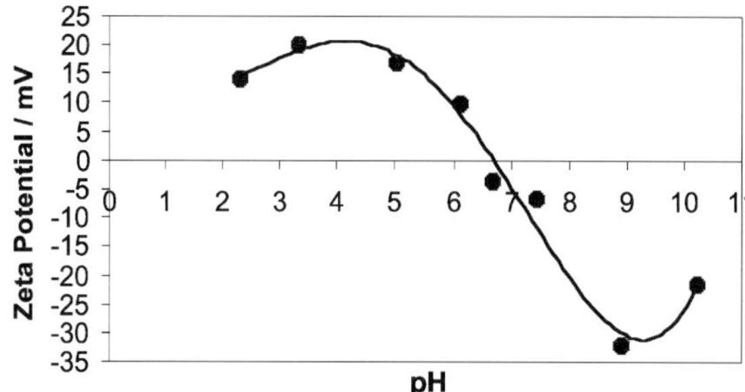

Fig. 1   Zeta potential of 1 vol.% PZT suspension as a function of pH.

possess a negatively charged double layer and above pH 6.5 thus becomes positive. In order for a suspension to be dispersed, replusive charges must be introduced onto the particle surfaces to counteract the attractive Van der Waals forces that always exist between particles.

Polyelectrolytes provide excellent stabilisation since they work using two different mechanisms, steric and electrostatic. Figure 2 shows the viscosity of PZT suspensions at 20 and 30 vol.% solids loading as a function of dispersant concentration. Viscosity measurements are commonly used to characterise the state of a dispersion. A minimum in viscosity depicts the optimum dispersant concentration. Therefore 1 wt.% D3201 is needed to stabilise the PZT particles at both 20 and 30 vol.% solids loading. Too little or an excess of dispersant present in a suspension may cause flocculation of the system and may also affect the overall viscosity of the suspension, consequently creating an unstable system.

Green tape characterisation

Figure 3 shows the tensile testing of tapes with various binder and plasticiser concentrations. It is expected as the vol.% PVA+PPG 400 decreases, the engineering tensile strength of the green tapes will decrease. 35 vol.% PVA+PPG 400 gave sufficient green strength for the handling of tapes required for further tape processing. Excessive amounts of binder and plasticiser present in the slip will push powder particles apart, which will produce detrimen-

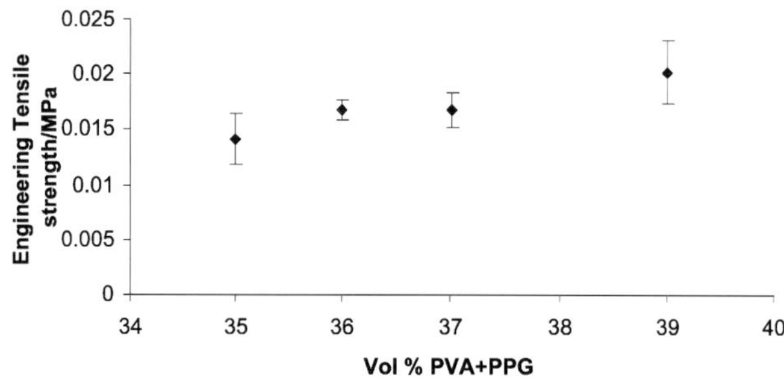

Fig. 2   Viscosity of 20 and 30 vol.% PZT suspensions as a function of D3201 amount.

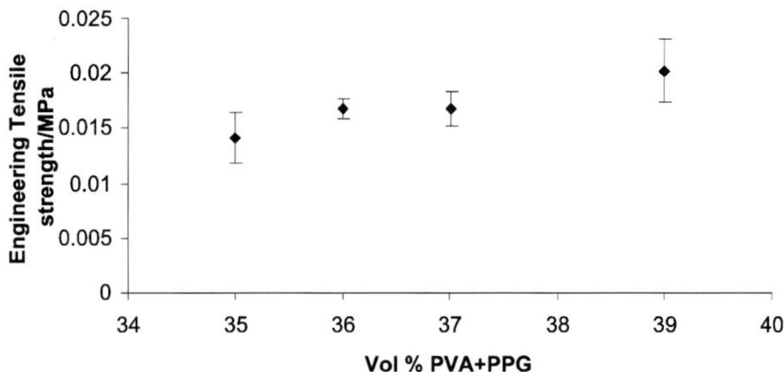

Fig. 3   Tensile strength of PZT green tapes with various PVA+PPG.

tal defects in the sintered ceramic. Also, the PZT solids content loading is reduced at higher organic concentrations producing lower green density tapes.

Figure 4 shows strain to failure of PZT green tapes with various PVA+PPG concentrations. The engineering strain to failure remains more or less constant to a value of 0.015 mm/ mm, when the binder and plasticiser concentration is reduced.

Dried tape thickness of green tapes

The dried tape thickness of green tapes are controlled by three factors: the slip's viscosity, the gap height and the casting rate during casting. For this experiment, the later two variables were kept constant.

The volume per cent water present in the slip will affect viscosity of the system. Figure 5. shows the dried green tape thickness as a function of the water content in the slip. The green thickness of tapes with 75.3 and 73.3 vol.% water ranged between 230–270 mm (upper limit). As tapes lose water during drying, they shrink both in the lateral and thickness directions. Therefore, it would not be practical to obtain tapes within the range 230–270 mm, when cast at 250 mm gap height. This indicates that the tape thickness was influenced by the viscosity of the system and not the gap height. In order to make tapes with good thickness control and high dimensional tolerance, the thickness of the tapes must be influenced by the

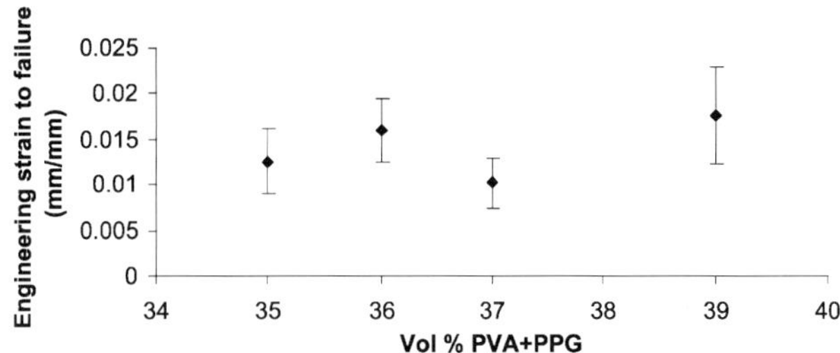

Fig. 4   Strain to failure of PZT tapes with various PVA+PPG concentrations.

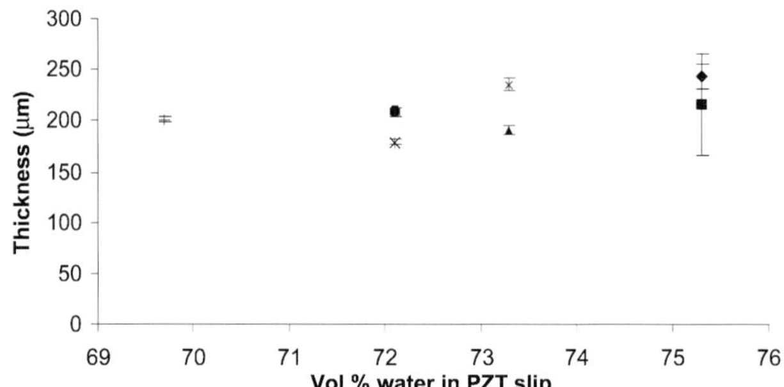

Fig. 5   Thickness of dries PZT green tape as a function of water content.

gap height and not the viscosity of the system. Therefore, slips with 69.7 and 72.2 vol.% water produced ≈200 mm tapes using a gap height of 250 mm. The variation in thickness along the tape could have been influenced by the change in wetting behaviour of the slips as the water content is varied.

From Table 1, it can be seen that slips with 72.2 and 69.7 vol.% water possess a viscosity of 0.15 and 0.27 Pa s^{-1}, respectively. These values may be needed for obtaining the green tape thickness for a gap height of 250 mm.

Table 1   Measured viscosity of slips 3 and 4 at a shear rate of 1000 s^{-1}.

Slip number	Vol.% water	Shear rate/s^{-1}	Viscosity/Pa s^{-1}
3	72.2	100	0.153
4	69.7	100	0.267

CONCLUSIONS

PZT suspensions were sucessfully stabilised using a polyelectrolyte by an electrosteric mechanism. The optimum dispersant level was obtained by viscosity measurements and correseponded to 1.0 wt.% D3201 based on PZT dry weight basis. The binder and plasticiser content was varied to increase the solids loading of PZT in the slip. Tensile testing was used to assess the green strength of the tapes. The solids loading of PZT improved from 57.4 to 63.0 vol.% PZT in the dried state.

ACKNOWLEGEMENTS

This work would not have been possible without financial support from Irisys Ltd. and EPSRC. R. W. Whatmore gratefully acknowledges the financial support of the Royal Academy of Engineering.

REFERENCES

1. R. E. Mistler, R. B. Runk and D. J. Shanefield, *Ceramic Processing Before Firing*, Wiley, 1978, 411–448.
2. E. P. Hyatt, *Ceram. Bull.*, 1996, **4**(65), 637.
3. R. E. Mistler, *Ceram. Bull.*, 1990, **6**(69), 1022.
4. D. Hotza and P. Griel, *Mater. Sci. Eng.*, 1995, **A202**, 206.
5. C. Gutierrez and R. Moreno, *J. Eur. Ceram. Soc.*, 2000, **20**, 1527.
6. C. Galassi, E. Roncari, C. Capiani and P. Pinasco, *J. Eur. Ceram. Soc.*, 1997, **17**, 367.
7. R. J. Mackinnon and J. B. Blum, *Adv. Ceram.*, 1984, **9**, 150.
8. T. Chartier and A. Bruneau, *J. Eur. Ceram. Soc.*, 1993, **12**, 243.
9. A. Kristofferson and E. Carlstrom, *J. Eur. Ceram. Soc.*, 1997, **17**, 289.
10. E. A. Groat, *Ceram. Ind.*, 1993, 34.
11. F. Dogan and J. H. Feng, *Mater. Sci. Eng.*, 2000, **A283**, 56.
12. M. C. B. Lopez, G. Fourlaris and F. L. Riley, *J. Eur. Ceram. Soc.*, 1998, **18**, 2183.
13. R. Moreno, *Am. Ceram. Soc. Bull.*, 1992, **71**, 1521.
14. R. Moreno, *Am. Ceram. Soc. Bull.*, 1992, **71**, 1647.
15. N. R. Gurak, P. L. Josty and R. J. Thompson, *Am. Ceram. Soc. Bull.*, 1987, **66**, 1495.
16. K. E. Burnfield and B. C. Peterson, *Ceram. Trans.*, 1992, **26**, 191.
17. S. L. Bassner and E. H. Klingenberg, *Am. Ceram. Soc. Bull.*, 1998, **77**, 71.
18. H. Tanaka, S. Fukai, N. Uchida, K. Uematsu, A. Sakamoto and Y. Nagao, *J. Am. Ceram. Soc.*, 1994, **77**, 3077.

Piezoelectric Transducer Arrays for Intravascular Ultrasound

E. L. Nix, S. C. Davies and K. M. Morel

Jomed Inc., Rancho Cordova, CA, USA

R. J. Dickinson

Jomed Imaging Ltd, Isleworth, UK

ABSTRACT

Techniques for the manufacture of miniature 64-element PZT transducer arrays for intravascular ultrasound are described. The arrays are cylindrical and less than 0.9 mm in overall diameter, comprising acoustic matching and backing layers and a central lumen. At these small dimensions the mechanical quality of the piezoelectric material is as important as its electrical performance. Some mechanical measurements on different PZT materials are presented to highlight key properties. Acoustic data is also presented to characterise the performance of the transducer elements and arrays.

INTRODUCTION

This paper presents a brief overview of some of the fabrication techniques employed in the manufacture of piezoelectric arrays used for intravascular ultrasound imaging, and a discussion of the piezoelectric materials themselves. Fixed multi-element arrays, rather than rotating single-element transducers, are preferred in clinical applications where the diagnostic imaging device is combined with a therapy, such as balloon angioplasty and stent placement. The geometry of coronary blood vessels dictates that the diagnostic fixed-arrays are cylindrical and, in general, less than 1 mm in diameter. A very small crossing-profile for the therapeutic 'combination catheter', which comprises imaging array, expanding balloon, stent and delivery system, is essential for access to small arterial lesions. The trend to create ever-smaller transducer arrays is technically challenging. Images have been obtained from 64-element arrays with an overall diameter as small as 0.85 mm, this dimension comprising an outer acoustic matching layer, a backing layer and a central lumen for a catheter guide-wire.

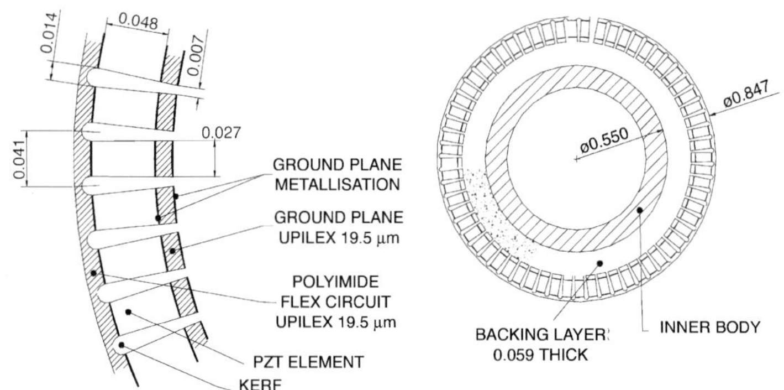

Fig. 1   Transducer configuration.

MINIATURE TRANSDUCER ARRAYS

To illustrate the tiny dimensions of the device as a whole, this section will describe the geometry of a 0.85 mm diameter transducer array. At the outset of the catheter build process the transducer consists of 48 μm thick PZT plate bonded between metallised polyimide layers, one layer (12.5 μm thick) being a grounding plane and the other layer (19.5 μm thick) being an acoustic matching layer, as illustrated in Fig. 1. Adhesive bonding is performed under controlled high pressure to ensure electrical continuity through sub-micrometre adhesive layers. Electrical grounding is achieved by folding the ground plane down to the matching layer ground-pad region and adhesive bonding them together. The whole assembly is then finely diced whilst in the flat configuration, before being formed into a precise cylinder by a wrapping process.

The acoustic matching layer is a high-resolution flexible circuit comprising 5 μm thick copper tracks on a 19.5 μm polyimide substrate. The circuit feature size is typically 10 μm, as described in more detail below. The copper tracks are the means of electrical address of the PZT array, the tracks being connected to multiplexing chips mounted on a second region of the flex circuit and themselves connected to the proximal end of the catheter. Dicing of the PZT using a 11 μm nickel–diamond blade yields 14 μm wide kerfs which also extend deeply into the polyimide substrate to give good acoustic channel isolation. The diced transducer element dimensions are 27 μm (width) by 48 μm (thickness) by 0.80 mm (length). At these tiny dimensions the mechanical quality of the piezoelectric material is as important as its electrical performance. Mechanical measurements will be presented to highlight key properties.

ARRAY DICING

Attention to process detail is critical for good dicing results at such miniature dimensions. A dicing platform with positional resolution of ± 0.1 μm and absolute positioning of ± 2 μm in each of the x, y and z axes is employed. The transducer assembly is mounted on a vacuum-chuck that is precision ground *in situ* on the dicing saw. Tight control of cutting spindle

Table 1 PZT material types.

Code	Origin	Composition	Grain size
GEC PNNZT-3	Special formulation prepared for Jomed by GEC Marconi, Caswell, UK	Solid solution of lead nickel niobate and lead zirconate titanate	3-6 µm
Aura C3900P	Aura Ceramics Inc., Minneapolis, USA	Unknown, possibly PMN-PZT	6 µm
ACL 4055 HP	Advanced Ceramics Ltd., Stafford, UK	PNNZT, similar to GEC	2-3 µm
TRS 200 FG HD	TRS Ceramics Inc. Pennsylvania, USA	Niobium doped PZT5A, said to be similar to Motorola 3195	0.5 µm claimed
TRS 600 FG HD	TRS Ceramics Inc. Pennsylvania, USA	Lanthanum doped PZT5H, said to be similar to Motorola 3203HD	1.0 µm claimed
PZN	TRS Ceramics Inc. Pennsylvania, USA	Flux-grown lead zinc niobate	Single crystal

speed (precision air-bearing with 60,000 rev min^{-1} capability), work-surface temperature ($\pm$ 0.5°C), blade-coolant temperature ($\pm$ 0.5°C), coolant flow rate and blade feed rate is essential for uniform and repeatable dicing results. In addition, accurate balancing of the nickel–diamond blade and a precisely defined blade-dressing procedure are critical for acceptable kerfs and undamaged, uniform array elements. Dicing process parameters will not be presented in this paper.

PZT MATERIALS

This section summarises an attempt made to apply the principles of fracture mechanics to the deformation and fracture of PZT 'planks' in three-point bend. The results provide a preliminary indication of the major differences between the PZT materials investigated and a tentative explanation of the differences in mechanical performance in miniature arrays. Table 1 lists some of the materials examined and their origins. The compositional information from suppliers is incomplete for commercial reasons.

For efficient transduction in the catheter ultrasound application, high values of both thickness-mode piezoelectric strain coefficient (d_{33}) and mechanical coupling factor (k_{33}) are desirable. Similarly a high value of relative permittivity (ε_r) is desirable because the miniature transducers are positioned at the end of a 2 m long transmission line. Moreover, for production-processing of the device at elevated temperatures, a high Curie transition temperature (T_C) is necessary. Unfortunately, this latter requirement conflicts with the need for high d_{33}. It is well-known that, in general, strongly piezoelectric PZT materials tend to have reduced T_C. Table 2 summarises the physical data available and illustrates the comments just made.

Modulus

Young's modulus, Y, was measured using a three-point bend technique with plank-shaped samples of dimensions 0.2 x 3 x 15 mm. The beam-centre deflection rate was ~1 mm s^{-1} and deformation was continued until fracture occurred. Simple beam-bending theory was

Table 2   PZT physical properties.

Property	Aura C3900P	GEC PNNZT-3	TRS 200 FG HD	TRS 600 FG HD	ACL 4055 HP	TRS PZN
d_{33} (pCN^{-1})	750	450	350	750	900	2200
ε_r (1kHz)	3900	4680	1810	3760	6000	5000
k_{33} (%)	78	71	68	78	80	94
T_C (°C)	195	173	345	190	155	185
Grain size (μm)	6	3-6	0.5	1	2	Single crystal
Density (kgm^{-3})	7820	8117	7700	7750	–	8200

employed, as in Timoshenko.[1] The instantaneous gradient of the plot of central load versus displacement gave Young's modulus at each value of the maximum strain in the stress distribution of three-point bend deformation. The results are summarised in Fig. 2. It is noted that the TRS materials (not shown) gave very similar results to the ACL samples, except that the maximum strain achieved for TRS 600 grade was only 0.21%. The significant variation of fracture strain amongst these materials is seen Table 3. The fracture strain of the single crystal PZN is reported to be 1.5% which would make this material very tough in practical applications.

Fracture

In the mechanical testing of materials, the fracture point is much less predictable than the modulus or the yield point. This is because fracture is related to the presence of flaws or

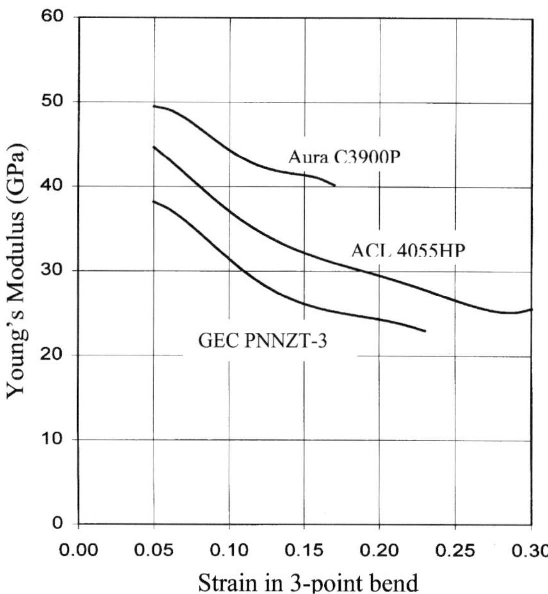

Fig. 2   OZT Young's modulii.

Table 3   PZT Mechanical properties.

PZT Material	Units	Aura C3900P	ACL 4055HP	GEC PNNNZT-3	TRS 200 FG HD	TRS 600 FG HD
Fracture strain	%	0.19	0.34	0.26	0.34	0.26
Weibull modulus, m	—	10.2	19.2	19.9	22.4	12.1
Weibull strength, σ_0	GPa	13.8	30.6	21.7	37.0	13.1
Intensity factor, K_C	MN/m$^{3/2}$	0.46	0.63	0.35	0.68	0.64
Surface energy, E	(J/m^2)	4.4	8.8	3.2	10.0	8.53
Intrinsic flaw size, a_i	μm	9.4	10.5	6.6	11.4	14.2
Thinning to 50μm	—	Success	Not done	Success	Success	Failure

defects whereas the extent of elastic and plastic deformation is a consequence of much more basic physical properties of the structure, such as inter-atomic forces. Ceramic materials fail by brittle fracture without measurable plastic deformation and consequently they show a wide distribution of failure strengths related to the presence of flaws. To characterise the behaviour of such materials it is usually necessary to quote failure probabilities. These can only be based upon the width of the distribution of fracture strengths for an adequate number of test samples of a particular composition. A simple statistical scheme based upon the three-point bend fracture test was employed, as in Broek[2] and Weibull,[3] the detail of which cannot be reproduced here for reasons of space. Sample sets were subjected to identical lapping processes and diced plank-shaped samples were of identical dimensions to permit valid comparisons. Figure 3 shows a typical 'Weibull plot' for the TRS 200 grade.

The significance of the statistical scheme is broadly as follows. Materials with a high Weibull modulus (e.g. $m > 20$) have a well-defined value of failure stress, because the risk of rupture, R, is defined by

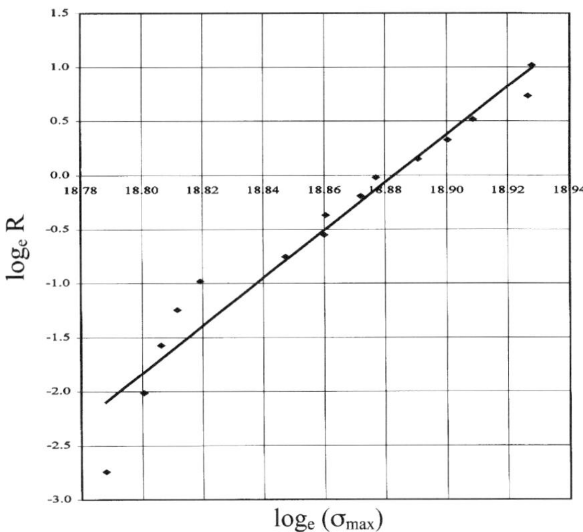

Fig. 3   Weibull data for TRS 200.

$$R = \iiint \frac{\sigma^m}{\sigma_0^m} dV_F \tag{1}$$

Here V_F is the stressed volume, $\sigma(x,y,z)$ is the stress distribution and σ_0 is a normalising constant related to strength. For high values of m, the risk R increases rapidly for stresses $\sigma > \sigma_0$ and, conversely, R decreases rapidly if $\sigma < \sigma_0$. For materials with low m the risk of rupture is spread over a wider range of σ values. That is, materials with low Weibull modulus have less predictable fracture strength and the design safety margin must be correspondingly greater. Thus m is as important as σ_0 in the discussion of fracture. A low value of m for PZT is considered to result from an increased concentration of intrinsic flaws. The Weibull modulus is equal to the gradient of the plot of $log_e R$ versus $log_e \sigma_{max}$ where σ_{max} is the maximum stress in three-point bend. It can be seen from Table 3 that the poor mechanical performance of the Aura material in dicing may be associated with its low Weibull modulus and low value of σ_0. The same comment applies to the TRS 600 grade, and, indeed, for this material we observed that it was exceedingly difficult to grind and lap to the 50 μm required for transducer assembly.

The success of these correlations encouraged a preliminary examination of additional material parameters. An estimate of the critical stress intensity factor, K_C, for crack propagation was made by creating artificial cracks in the PZT. That is, notches of depths 15 to 55 μm were cut into the underside centre of the three-point bend specimens. The analytical details are set out in Broek.[2] In broad terms, the three-point bend fracture load, F_C, is related to notch depth, a, by

$$F_C = K_C \frac{wt^{3/2}}{2LG(a/t)} \tag{2}$$

where w and t are the plank width and thickness, $2L$ is the plank length between bend points and G is a polynomial function of a/t, the ratio of notch depth to plank thickness. G can be regarded as a geometrical factor. A plot of F_C versus $wt^{3/2}/(2LG)$ is a straight line through the origin with gradient K_C (since F_C is zero for very large L). Figure 4 shows such a plot for Aura C3900P. Once an average value for K_C has been determined using a range of notch depths, the inherent flaw size, a_i, can be estimated from

$$a_i = \frac{1}{\pi} \left(\frac{wt^2 K_C}{3F_C L} \right)^2 \tag{3}$$

where, here, F_C is the average fracture load for un-notched specimens.

It was hoped that there might be a simple relation between the inherent flaw size and the material grain size seen in SEM examinations. However, no clear correlation was observed, as seen in Table 3. It seems reasonable that material cohesion is represented by both stress intensity factor and surface energy, E, the latter being calculable from K_C according to $E = (K_C)^2/Y$. The materials study is summarised in the conclusion.

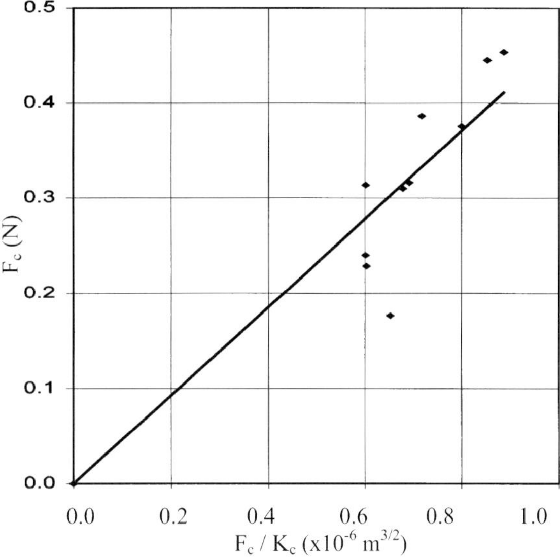

Fig. 4   Intrinsic flaw size estimation.

MINIATURE FLEXIBLE CIRCUITS

The requirement for a small diameter 64-element PZT array leads to the necessity for a miniature circuit onto which this array can be bonded. Typical track and gap dimensions on the circuits used in our catheters are 10 and 13 μm respectively, over a length of approximately 20 mm, although circuits of up to 80 mm in length have been produced. The thickness of these tracks over the majority of this length is approximately 5 μm, whilst in the area of the PZT array this is reduced to 1.5 ± 0.5 μm to aid acoustic matching, as illustrated in Fig. 5.

Fig. 5   The transition region between the 5 and 1.5 μm tracks. The PZT array region is to the right of the figure. The edges of the tracks, and the gap between them (centre of figure), can be seen clearly.

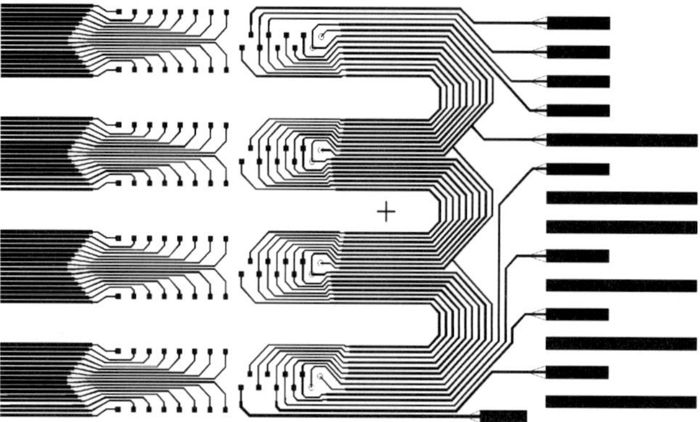

Fig. 6   A diagram of the integrated circuit bond pad region. The 26 copper bondpads are to the left of the figure, whilst the region where the 14-way ribbon cable is soldered to the copper circuits is on the right.

The tracks are made of copper, which is used because of its good electrical characteristics, as well as its ability to form good solder joints using traditional tin–lead technology. These solder joints are formed between each miniature circuit and the integrated circuit chips used to control the function of the PZT array. Additionally solder joints are formed between the flex circuit and the ribbon cable used to carry the electrical signals along the entire length of the catheter.

The control integrated circuits are essentially multiplexing switches, of dimensions 0.1 mm thickness by 0.5 mm width by 1.5 mm length. There are typically four of these per device, each of which has 26 or more Sn–Pb solder bumps which have to be aligned to within ± 5 µm of the 35 µm square copper bond pads, as illustrated in Fig. 6.

The miniature circuits are manufactured using standard photo-lithographic techniques, i.e. pre-baking, photo-resist coating, resist baking, photo-mask alignment and contact exposure, development, copper electroplating and final etching, with these final two stages being 'wet' chemical processes. The photo-mask has a positive polarity, so that the photo-resist is removed in the areas where the copper tracks are required, and the copper then built up in these areas during the electroplating. An additive, rather than subtractive, process is used because of the large track thickness-to-width ratio and the need for well-defined, straight edges at these track and gap dimensions.

Since a circular cross section for the PZT array is required in this application, these copper circuits are made on a flexible 20 µm polyimide substrate that is then wrapped into a cylindrical shape. This has the advantage that the polyimide forms the matching layer between the PZT elements and the blood, but the disadvantage that the movement of the polyimide during the circuit manufacture can limit track resolution and hence yield. Despite this, yields of up to 75% perfect circuits have been achieved with our latest design.

ACOUSTIC CHARACTERISATION

The machining of piezoelectric materials to form the miniature arrays required for intravascular ultrasound is inherently problematic. The dicing process exerts stresses on the array, which can result in chipping, fracturing and crumbling of the individual array elements. In addition there is a piezoelectric dead-zone a few micrometres thick adjacent to the dicing blade. In order to be able to assess inter-element uniformity and individual element acoustic response it is necessary to perform acoustic measurements on the machined arrays.

A measure of inter-element uniformity is achieved by performing pulse-echo measurements on each element in turn. The test equipment for this measurement comprises a flat specular metal target positioned 2 mm from the surface of the front face of the transducer array. The propagating medium between the transducer and the target is water. Individual elements are fired with a four cycle tone burst with a typical centre frequency of 25 MHz. An oscilloscope is used to monitor the echo response for each element. The measure of inter-element uniformity is derived from a comparison of the time-averaged amplitudes of the echo responses.

This method of uniformity assessment is simple enough to be used in routine process monitoring. A disadvantage is the sensitivity of the echo to the correct positioning of transducer array with respect to the target. Misalignment or flexing of the array will yield erroneous results. Fixtures were developed to facilitate rapid positioning with minimal risk of misalignment. Damage to the PZT manifests itself as a drop in the amplitude of the echo response and often occurs in four or five adjacent elements.

Inter-element uniformity measurements are also performed after the array has been formed into the cylindrical transducer. The method is similar to that described for the flat array except that the target comprises a cylindrical lumen into which the transducer is inserted. The inner diameter of the lumen is 4 mm greater than the external diameter of the transducer allowing a transducer-target separation of 2 mm. Again the alignment of the transducer with respect to the target is critical; misalignment results in artefactual trends in amplitude superimposed on the measurements. The measurements performed on the cylindrical transducer indicate whether elements have been damaged in the wrapping process and also give an indication of the concentricity of the device. Other transducer build faults can also be detected from the uniformity results. These defects include voids in the backing layer, delamination of the piezoelectric material from the base polyimide and electrical shorting of adjacent elements.

The acoustic characteristics of individual elements are obtained by generating a two-way transfer function. This is achieved by scanning a fine wire target (50 μm diameter) in a circle of ~2 mm radius around a transducer element and obtaining a time-averaged pulse-echo response for selected angular positions. This automated measurement is performed for elements on the flat array and also elements in the cylindrically formed device. The two-way transfer function gives an indication of the acoustic beam shape formed by a transducer element. Theory suggests that for a hexahedron-shaped element the two-way transfer

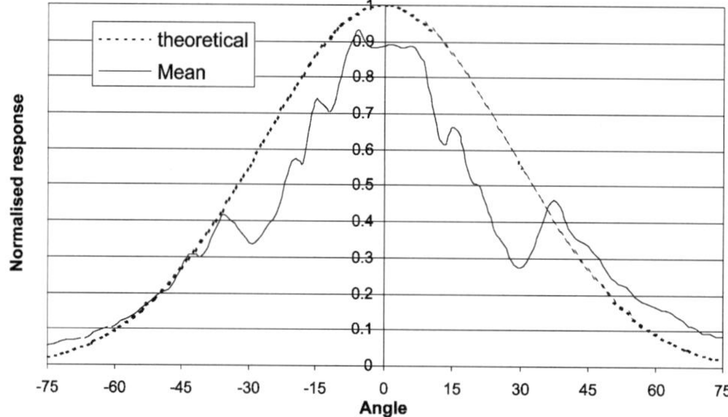

Fig. 7   Mean two-way transfer function for a transducer array.

function should describe a sinc function. The plot shown in Fig. 7 represents the mean two-way transfer function obtained from a transducer array.

Inspection of Fig. 7 shows that the overall shape of the measured two-way transfer function is close to that predicted by theory, namely a sinc function. It can also be seen that there are side lobes present at approximately $+40°$ and $-40°$. The side lobes are a consequence of acoustic cross-talk between adjacent elements. Increased cross-talk gives more prominent side lobes. Cross-talk is primarily a consequence of the mechanical linkage between adjacent elements and can occur through the front face, the backing layer or the kerf-fill material (if the latter is present). It has been shown that cross-talk can be significantly reduced by dicing through both the piezoelectric material and the base substrate. In practice, of course, it is not feasible to dice all the way through the substrate. A direct consequence is that it is necessary to control the dicing height with great precision so that partial cuts are made into the polyimide leaving only a few micrometres of bridging material remaining. Cross-talk can also be suppressed by removing any material from the kerfs and by ensuring that there are no bridges of adhesive between adjacent elements.

IMAGE QUALITY

For the intravascular ultrasound device described in this paper, imaging is achieved using a synthetic aperture. As a consequence the image quality is dependent on both the inter-element uniformity and the shape of the two-way transfer function. The greater the variation in the inter-element responses the poorer the image quality. However, by increasing the width of the aperture it is possible to reduce the dependency on uniformity.

To obtain the best image quality it is necessary for the two-way transfer function to be similar in shape to the two-way transfer function predicted by theory. The presence of side lobes leads to artefacts in the reconstructed image. Similarly, if the main lobe is too narrow, image quality is compromised.

MATERIALS SUMMARY

PZT materials selection is an ongoing study because of the significant impact of PZT mechanical quality on the final image, and on product cost. Interim conclusions are presented here. In respect of the promising fine-grain materials, our device fabrication dilemma can be summarised as follows. TRS 200 has excellent mechanical properties but inadequate electrical properties (the latter conclusion based on flat-array pulse-echo results) whereas the TRS 600 grade has good electrical properties but poor mechanical properties. That is, the 600 grade was exceedingly difficult to lap into 50 µm sheets, and bend-tests on 0.2 mm thick planks showed significantly inferior and variable mechanical properties for the 600 in comparison with the 200 grade. We therefore abandoned both of these materials. Of the remaining materials it is notable that Aura C3900P had the highest modulus but the lowest strain value at failure and, in addition, an inferior Weibull modulus and strength. This data was also supported by the poor dicing results with Aura. Work with Aura was therefore discontinued. ACL 4055HP showed a high stress-intensity factor and fracture-surface energy, which, it was tentatively concluded, contributed to the high fracture strain of this material. However, the intrinsic flaw size appears to be similar to the Aura, which might be expected on the basis of the similar grain size. It is noted that the high d_{33} and high permittivity of the ACL material make it an attractive candidate for the intravascular transducer application.

GEC PNNZT-3 is a low modulus material compared with the Aura 3900P. However, extensive SEM examinations of sectioned arrays showed that the GEC material gave dicing results markedly superior to those obtained with the Aura. A tentative conclusion is that this result is consistent with the superior fracture strain values, and with the smaller intrinsic flaw size, of the GEC compared with the Aura. It is understood that there may be a geometrical scaling factor in the estimation of the intrinsic flaw size.

No fracture data was gathered for the single crystal PZN because of the high cost of procuring samples and the foreboding that PZN arrays would never be commercially viable. However, preliminary fine-pitch dicing of 0.6 mm thick disc samples in our possession demonstrated excellent kerf-wall properties under SEM examination. If the cost of such materials can be reduced to commercially acceptable levels for disposable devices then single crystals become front-running contenders for intravascular applications because of their astonishingly high d_{33} and k_{33} values.

ACKNOWLEDGEMENTS

The work described above was performed at Jomed Imaging Ltd, Isleworth, UK. The project was transferred to Jomed Inc., Rancho Cordova, California, USA, during March 2001. The authors wish to thank Jomed Inc. for permission to publish.

REFERENCES

1. S. Timoshenko and J. N. Goodier, *Theory of Elasticity*, 2nd edn, McGraw-Hill, 1951, 99–101.
2. D. Broek, *Elementary Engineering Fracture Mechanics*, Sijthoff & Noordhoff, 1978.
3. W. Weibull, 'A statistical distribution function of wide applicability', *J. Appl. Mech.*, 1951, **18**, 293.

Preparation and Properties of BSTO Thick Films on Metal Substrates

P. Kr. Petrov, V. Sarma, M. Poole and N. McN. Alford

Centre for Physical Electronics and Materials, School of Electrical, Electronic and Information Engineering, South Bank University, 103 Borough Road, London SE1 0AA, UK

ABSTRACT

$Ba_xSr_{1-x}TiO_3$ thick films with different thickness and varying Ba/Sr ratios were sintered on silver substrates at 900°C. Two powder routes were tested: mixtures of $BaCO_3$, $SrCO_3$ and TiO_2 and mixtures of $SrTiO_3$ and $BaTiO_3$ powders. The structural properties of the powder mixtures and sintered films were investigated using SEM and X-ray diffraction. The RF electrical properties were examined at 1 MHz in the temperature range between 260 and 320 K. Even with a significant degree of porosity (50–60%), the BSTO thick films were almost mono-phase and exhibited electrical tuning 5–10% under electrical field of about ~5 kV cm^{-1} (1 MHz, 300 K), and Q-factors of about 80–110 (1 MHz, 300K).

INTRODUCTION

Ferroelectric thick films of $SrTiO_3$ (STO) and $BaTiO_3$ (BTO) and their solid solution $Ba_xSr_{1-x}TiO_3$ (BSTO) are the subject of considerable research and show some promise in tuneable microwave devices.[1–3] Despite the comparatively inferior electrical properties (high loss level, low tuning ratio), thick non-epitaxial BSTO films made by ceramics technology are still attractive for many applications because of their low fabrication cost and good reproducibility.[4,5] Usually, dielectric materials are used as substrates for such thick films, although, for many applications good heat removal and conductive bottom electrodes are needed. These requirements may be satisfied using metal substrates.

In this paper, a set of BSTO thick films with different thickness and varying Ba/Sr ratios were sintered on silver substrates at the relatively low temperature of 900°C. Their structural properties were investigated using SEM and X-ray diffraction. The RF electrical properties were examined at 1 MHz in the temperature range between 260 and 320 K.

EXPERIMENTAL

As the intention of our work is to obtain a device operating at room temperature, initial BSTO powders were prepared with Ba/Sr ratios of both 0.75:0.25, $(Ba_{0.75},Sr_{0.25})TiO_3$, and

0.5:0.5, $(Ba_{0.5},Sr_{0.5})TiO_3$, using two methods. Method 1 is a mixture (with the appropriate Ba/Sr ratio) of $BaCO_3$, $SrCO_3$ and TiO_2 source powders, and method 2 is a mixture of $BaTiO_3$ and $SrTiO_3$ powders.

The powders were mixed for 24 h using 10 mm zirconia grinding media and then were wet milled in H_2O for 240 h to improve homogeneity. After drying at 80°C, the resultant cakes were ground in a mortar and pestle and passed through a 40 μm sieve. After calcination (2 h at 1100°C) powders were dry ball milled for another 240 h and then the surface area was measured using a Coulter 3100SA (Coulter Electronics).

Thick film inks using the powders made from the two different methods were prepared and deposited onto silver discs (20 mm diameter, 1 mm thickness) and dried for 1 h at 80°C. The pre-sintered thickness was approximately 100 μm. The thick films were sintered in air for 6 h at 900°C on Ag substrates.

The structural properties of all samples fabricated were investigated using SEM (Hitachi S-800 FEG) and X-ray diffraction (Philips X'Pert). The RF electrical properties were measured in a specially designed test-fixture. The measurements were carried out at 1 MHz using HP LRC-meter (HP 4263B) in the temperature range between 260 and 320 K. For these temperature measurements, the test-fixture was anchored onto a Peltier element and controlled by a Lake Shore 330 temperature controller. To exclude the parasitic capacitance and resistance introduced by the test fixture, short and open circuit calibrations were performed before measurements.

RESULTS AND DISCUSSION

The results of SEM analysis are presented in Fig. 1. Samples made by both methods exhibited a high degree of porosity. The porosity was calculated by assuming a theoretical density

Fig. 1   Typical SEM images of thick $(Ba_{0.5},Sr_{0.5})TiO_3$ films made from: (a) mixtures of $(BaCO_3$, $SrCo_3$ and $TiO_2)$ powders; (b) mixtures of $(SrTiO_3$ and $BaTiO_3)$ powders.

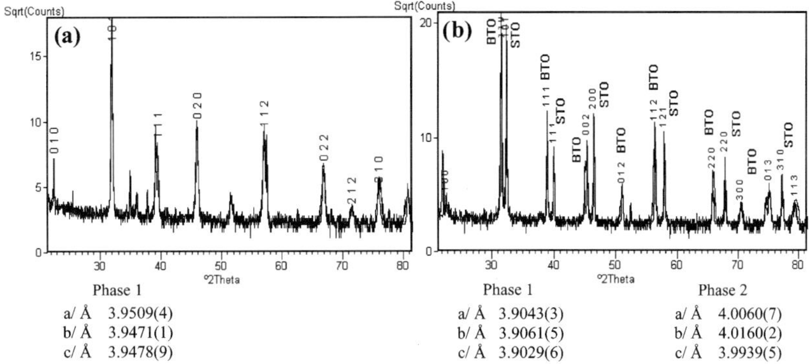

Fig. 2   Typical XRD images of thick $(Ba_{0.5},Sr_{0.5})TiO_3$ films made from: (a) mixtures of $(BaCO_3, SrCo_3$ and $TiO_2)$ powders; (b) mixtures of $(SrTiO_3$ and $BaTiO_3)$ powders.

for $SrTiO_3$ of 5.119 g cm^{-3}, and for $BaTiO_3$ 6.02 g cm^{-3}, and was found to be between 50 and 60%.

Figure 2 shows typical results of θ–2θ XRD measurements for the samples prepared by both method 1 and method 2. Even with high porosity, the samples made by method 1 (Fig. 2a) have one well-presented BSTO phase and almost no secondary phases. The BSTO phase is absent in all samples made by method 2 (Fig. 2b) while the peaks from STO and BTO phases are clearly evident. This fact explains the temperature behaviour of the capacitance (Fig. 3). Sample 1 had a well-defined peak at 296 K (corresponding to the Curie temperature phase-transition,) typical for $(Ba_{0.5}Sr_{0.5})TiO_3$ materials (Fig. 3a).

The broad peak in the C–T dependence of sample 2 (Fig. 3b) is probably a result of the existence of two phases.

The experimental set-up allowed an upper limit of 40 V dc bias applied on the sample. The resulting electrical field of about ~5 kV cm^{-1} is too small in comparison with one usually

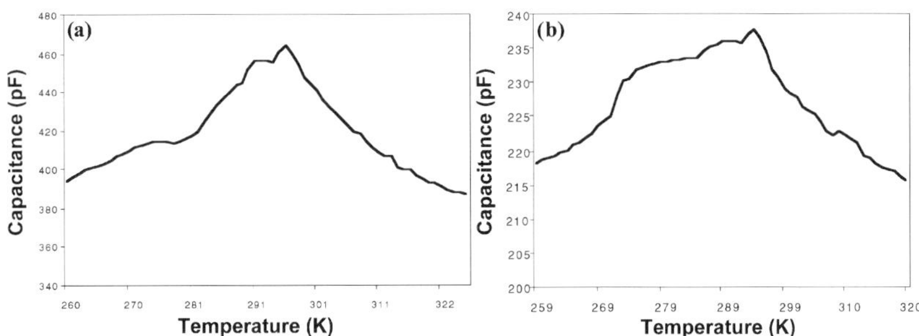

Fig. 3   Temperature dependence of thick $(Ba_{0.5},Sr_{0.5})TiO_3$ films measured at 1 MHz: (a) sample made by mixtures of $(BaCO_3, SrCo_3$ and $TiO_2)$ powders; (b) sample made by mixtures of $(SrTiO_3$ and $BaTiO_3)$ powders.

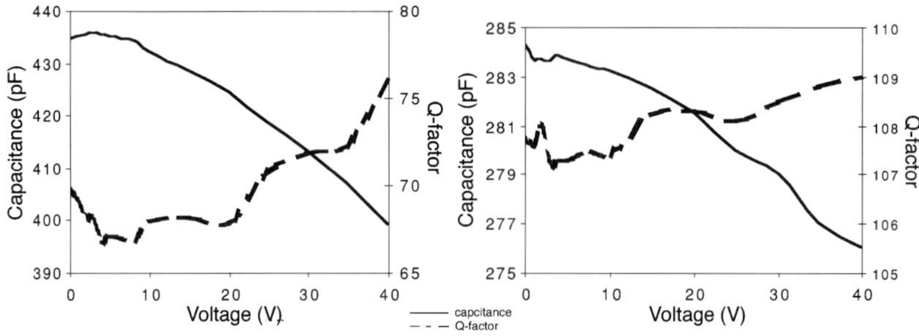

Fig. 4   Electrical properties of (a) $Ba_{0.75}Sr_{0.25}TiO_3$ and (b) $Ba_{0.5}Sr_{0.5}TiO_3$ films made by mixtures of $(BaCo_3, SrCO_3$ and TiO_2 powders (method 1).

applied for electrical tuning,[6] therefore no tuning was detected for the samples made by mixture 2. Nevertheless an electrical tuning ratio [C(0V)-C(350V)]/C(0V) of about 10% and Q-factor about 70 was measured for a sample made by mixture 1 with 75:25 Ba/Sr ratio (Fig. 4a). The Q-factor measured for the $(Ba_{0.5}Sr_{0.5})TiO_3$ sample was above 100, while the tuning ratio was about ~2-3% (Fig. 4b).

The low Q-factor and tuning ratio might be explained with the 50–60% porosity of the BSTO films, which reduced the effective ε_r and hence the tuning capability. However, we do not yet have a good explanation why mixture 2 is two phase, while mixture 1 was single phase, particularly as the surface area of the pre-calcined powder used for mixture 1 (about 0.7 m^2 g^{-1}) was almost four times smaller than the surface area measured for mixture 2 (about 3.0 m^2 g^{-1}). The only speculation, which we have, is that the solid-state diffusion between the elements in mixture 1 is faster. This problem is under current investigation and will be the subject of further research.

CONCLUSIONS

In this paper, a set of $Ba_xSr_{1-x}TiO_3$ thick films with different thickness and varying Ba/Sr ratios were sintered on silver substrates at a relatively low temperature of 900°C. Two powder routes were tested: mixtures of $BaCO_3$, $SrCO_3$ and TiO_2 and mixtures of $SrTiO_3$ and $BaTiO_3$ powders. The results show that powder mixture 1 is preferable for low temperature sintering. Even with large amounts of porosity (50–60%), the thick BSTO films made with this mixture were almost mono phase and exhibited electrical tuning of 5–10% under electrical field of about ~5 kV cm^{-1} (1 MHz, 300 K), and Q-factors of about 80–110 (1 MHz, 300 K). Experiments are in progress to decrease the porosity in films by using high temperature metal/alloy substrates as well as alternative film preparation techniques (such as sol–gel, pulsed laser deposition). This is expected to improve the electrical properties and in particular the Q-factor and tuning.

ACKNOWLEDGEMENTS

This work was partly supported by the Engineering and Physical Sciences Research Council and Framework V (FOAM).

REFERENCES

1. P. K. Petrov, Z. G. Ivanov and S. S. Gevorgyan, EuMC'2000 Conference Proceedings, 2000, Vol. 3, 218–221.
2. M. J. Lancaster, J. Powell and A. Porch, *Supercond. Sci. Technol.*, 1998, **11**(11), 1323.
3. R. W. Whatmore, *Ferroelectrics*, 1999, **225**(1/4), 985.
4. J. E. Holmes, D. Pearce and T. W. Button, *J. Eur. Ceram. Soc.*, 2000, **20**(16), 2701.
5. A. W. Tavernor, H. P. S. Li, A. J. Bell and R. Stevens, *J. Eur. Ceram. Soc.*, 1999, **19**(9), 1691.
6. P. K. Petrov, E. F. Carlsson, P. Larsson, M. Friesel and Z. G. Ivanov, *J. Appl. Phys.*, 1998, **84**(6), 3134.

Synthesis of $0.94(Bi_{1/2}Na_{1/2})TiO_3$–$0.06BaTiO_3$ by Spray-Drying

A. Sanson

Advanced Materials Group, School of Industrial and Manufacturing Science, Cranfield University, Cranfield, Bedford MK 0AL, UK

C. Galassi and A. L. Costa

CNR-IRTEC, Research Institute for Ceramics Technology, National Research Council, Faenza, Italy

U. Russo

Department of Inorganic Metallorganic and Analytical Chemistry, University of Padua, Italy

ABSTRACT

The solid solution between bismuth sodium titanate, $(Na_{1/2}Bi_{1/2})TiO_3$, and barium titanate, $BaTiO_3$, is considered to be a good candidate for producing lead-free piezoelectric materials. Spray-drying a solution of salts or metal organic compounds is a more flexible process compared with conventional solid state reaction of the oxides and can allow better control of the stoichiometry and a reduction in the temperature of perovskite phase formation. In this study, powders of composition $0.94(Bi_{1/2}Na_{1/2})TiO_3$–$0.06BaTiO_3$ were prepared by spray drying of an aqueous solution of $Bi(NO_3)_3.5H_2O$, $Ba(CH_3COO)_2$, $NaNO_3$ and $Ti[OCH(CH_3)_2]_4$, followed by thermal treatment and characterised by XRD, SEM, BET and AAS. Essentially spherical and hollow aggregates of highly reactive sub-micrometre particles were obtained that allowed the calcination temperature to be reduced by 150°C compared with powders prepared by the conventional mixed oxide route.

INTRODUCTION

Among piezoelectric ceramics, lead-based perovskites exhibit the best piezoelectric properties and are widely used in several practical applications. However, the high toxicity of lead oxide and its high vapour pressure during processing has increased interest towards more environmentally compatible materials. In the search for a lead-free system capable of showing good piezoelectric properties considerable attention has been paid to the $(Na_{1/2}Bi_{1/2})TiO_3$ system and its solid solutions and especially the $BaTiO_3$–$(Na_{1/2}\,Bi_{1/2})TiO_3$ solid solution system (BNBT). In particular, the composition corresponding to the rhombohedral-tetragonal morphotropic phase boundary ($0.94(Bi_{1/2}Na_{1/2})TiO_3$–$0.06BaTiO_3$) exhibits the best piezoelectric properties found to date in ceramics without lead.[1,2]

The physical and chemical characteristics of the precursor powder strongly influence the final electrical and mechanical properties of the ceramics. In particular, the purity, size distribution and reactivity of the precursor powder are very important and should be carefully controlled. Unlike conventional solid state reaction techniques, methods that use solutions (sol–gel, coprecipitation of hydroxides or oxalates, spray-pyrolisis, freeze-drying and spray-drying) can lead to very homogeneous, fine (sometimes nanosized) powders that require lower temperature and time of calcinations.[3] All these techniques produce precursors with very precise stoichiometry and well determined physical and chemical properties. Among these spray drying is a relatively simple and cheap process which leads to close control of the stoichiometry of the powders.[4] The spray-drying process can be divided in four principal phases:

- atomisation of the solution in droplets
- mixing of the spray with the air flowing in the hot chamber
- evaporation and drying of the droplet
- separation of the dried particles from the air flow.

All these steps contribute to the final morphology of the particles and must be controlled. In fact, it is very important to produce tiny and dense particles, to obtain compounds with good mechanical and piezoelectric properties.[5]

The aim of the work presented here is to obtain fully reacted powders from the system $0.94(Bi_{1/2}Na_{1/2})TiO_3$–$0.06BaTiO_3$ by spray-drying and to analyse their characteristics with powders obtained by the mixed oxide route.

EXPERIMENTAL

Powders of the composition $0.94(Bi_{1/2}Na_{1/2})TiO_3$–$0.06BaTiO_3$ (BNBT) were prepared either by mechanical mixing of the oxides (route MM) or by spray-drying of the solution of the precursors (route SD). For the MM route the stoichiometric amounts of oxides powders (Table 1) were ball milled in water for 48 h using zirconia balls as milling media. Water was removed by freeze-drying (Edwards MFD 0,1) and the powder sieved at 400 μm. The MM powder was then calcined at 800°C, ground, sieved and granulated by humidification with an aqueous solution at 3 wt.% PEG. In SD, the solution of the precursors (Table 2) is sprayed

Table 1   Raw materials for mechanical mixing (MM).

Reagent	Producer	Purity, %
$BaCO_3$	Merck	99.4
TiO_2	Degussa	99.4
Na_2CO_3	Merck	99.5
Bi_2O_3	Aldrich	99.9

Table 2   Raw materials for spray-drying (SD).

Reagent	Producer	Purity, %
$Bi(NO_3)_3.5H_2O$	Carlo Erba	98.5
$NaNO_3$	Carlo Erba	99.5
$Ti(Oi-Pr)_4$	Aldrich	97.0
$Ba(Ac)_2$	Merck	99.0

in a hot air stream producing spherical granules. Bismuth nitrate, sodium nitrate, barium acetate and titanium isopropoxide are dissolved under stirring in nitric acid in presence of hydrogen peroxide. The as-prepared solution is sprayed and dried in A Labplant SD-05 spray drier (air stream temperature 220°C, air flow 66 $m^3\,h^{-1}$, pump suction flow 168 $cm^3\,h^{-1}$). The powder is then calcined (650°C), ground, sieved and granulated in the same way as the MM powder.

The purity of the raw materials and stoichiometry of the as-prepared powders was determined by Inductively Coupled Plasma-Optical Emission Spectrometry (Spectrometer Varian, Mod. Liberty 200). Thermogravimetry (TGA) and differential thermal analysis (DTA) of the dried powders were carried out at a heating rate of 20 K min^{-1} (Simultaneous Thermal Analyser STA 409 Netzsch). The crystalline phases were identified by powder X-ray diffraction (XRD) (Rigaku, Mod. Miniflex) operating with Ni-filtered Cu K_α radiation over the 2θ range between 20 and 60 degrees.

Particle size distribution was measured by a Sedigraph 5100 (Micromeritics) and specific surface area by BET single point method (Sorpty 1750 Carlo Erba); the morphology of the powder was observed by SEM (Leica Cambridge Stereoscan 360). Apparent density was measured by the Archimede's method in water.

RESULTS AND DISCUSSION

Preparation of the solution

In order to obtain the precursor powder it is necessary to dissolve the precursor salts that will lead to the system of interest: the formation of the solution represents a key step of the process. Aqueous solutions are the most common used because they are easily handled, safe, cheap and above all, there are many water soluble salts.

In general, chloride and oxychloride are the most soluble compounds but the corrosive nature of the gases that derive from them and the bad effect of chlorine on the sintering process, have driven towards the use of less soluble but less dangerous salts, such as nitrates, acetates and sulphates. Titanium is the most difficult cation to dissolve: all its salts are insoluble in a neutral aqueous environment and is only sparingly soluble in an acid environment. The most soluble compound on sale is $Ti[OCH(CH_3)_2]_4$, nevertheless it presents a significant solubility only in a strongly acidic environment (pH $\leq$ 0). Free Ti^{4+} doesn't exist in water, but forms hydrates containing the $Ti(OH)^{3+}$ and $Ti(OH)_2^{2+}$ groups that in very acidic environment (pH $\approx$ 0) are in equilibrium with TiO^{2+} (Ref. 6). At slightly higher values of pH, species such as $Ti_3O_4^{4+}$ are present and can lead to the formation of colloids or to precipitation of $TiO_2.nH_2O$. Titanium hydroxide $Ti(OH)_4$ precipitates at pH = 1: raising the temperature or diluting the solution changes it into less soluble alkaline titanates so it is not possible to promote solubility by increasing the temperature. Hydrogen peroxide increases the stability in solution of Ti^{4+} forming the orange peroxycomplex TiO^{2+}, $[Ti(O)_2OH_{(aq)}]$; in this way the free titanium IV is less available. With regard to bismuth, at pH = 0 the cation exists free in the 3+ oxidation state. In the presence of nitrates it forms complexes of the form $Bi(NO_3)_4-n(H_2O)_{2n}^{2+}-Bi(NO_3)_4^-$ (Ref. 7). The dilution, even local, of a strongly acidic solution of Bi in HNO_3 results in precipitation of $BiO(NO_3)$ and $Bi_2O_2(OH)(NO_3)$. The behaviour of the salts of sodium and barium are governed by simple solubility equilibria. The analysed system is quite complex, because of the large number of equilibria involved (of unknown constants) and of the mutual interference of the cation in solution. Therefore it was not possible to establish the theoretical solubility of the salts. Various tests at increasing HNO_3 concentration (1 M, 1.5 M, 2 M, 2.5 M, 3 M) were then performed to determine: the highest concentration of salts in solution; the less aggressive environment; the best stability as a function of time. The final solution obtained is indicated in Table 3.

SPRAY DRYING TECHNIQUE

The process conditions were optimised to obtain dense particles and good piezoelectric and mechanical properties. To improve the homogeneous volumetric precipitation, apart from a high concentration of the salts in solution, it is necessary to have droplets as small as possible and with a low speed in the hot chamber. In this way, the evaporation and consequent precipitation are gradual and homogeneous all over the droplet. The solution was atomised with a two fluid pneumatic co-current system. In order to ensure a small particle size distribution it is necessary to use a high pressure of the atomising fluid. In doing so the droplets gain high kinetic energy and very quickly pass over the hot chamber: it is then necessary to have a high temperature in the chamber to induce the precipitation of the salts in the droplets. Moreover, a high speed can cause the particles to collide with the wall of the chamber before the process of drying is complete. It follows that the process parameters to optimise are:

Table 3   Composition of the sprayed solution.

Compound	Concentration
$Ba(CH_3COO)_2$	0.01 M
$Bi(NO_3)\cdot 5H_2O$	0.100 M
$NaNO_3$	0.100 M
$Ti[OCH(CH_3)_2]_4$	0.214 M
HNO_3	1 M
H_2O_2	0.01 M

Table 4   Range of the process parameters and final values used.

Parameter	Range of optimisation	Final value
Pump flow (cm^3/h)	168-350	168
Airflow (m^3/h)	62-66	66
Atomising air pressure (bar)	1,9-2,1	2,1
Nozzle diameter (mm)	0,5; 0,7; 0,9	0,5
Chamber temperature (°C)	200-220	220

- flow of the pumped solution
- pressure of the atomising air
- airflow that drives the droplets to the chamber
- chamber temperature.

The ranges used for the optimisation and the final values of the parameters are summarised in Table. 4.

Powder characteristics

From the chemical analysis the stoichiometry of the powders corresponds to the nominal value, with differences lying within the experimental error.

The thermal analysis shows that the spray-dried powder had reacted fully below 650°C (Fig. 1). By 200°C the powder had already lost 20% of its initial weight. This initial weight loss corresponds to the loss of absorbed water of the organic part of the titanium isopropoxide and of the nitrates. There is a further 15% loss when the sample is heated from 200 to 600°C. Above that temperature the thermal process seemed to be finished. The peak at 640–659°C should be the one referred to the formation of the solid solution of interest. The X-ray diffraction pattern of the powder calcined at 350–650°C (Fig. 2) confirms the transformation into the perovskitic rhombohedral phase. The system passes through the formation of the more stable pyrochlore phase, the most intense peaks ($2\theta = 30$) being observed at ~350°C in

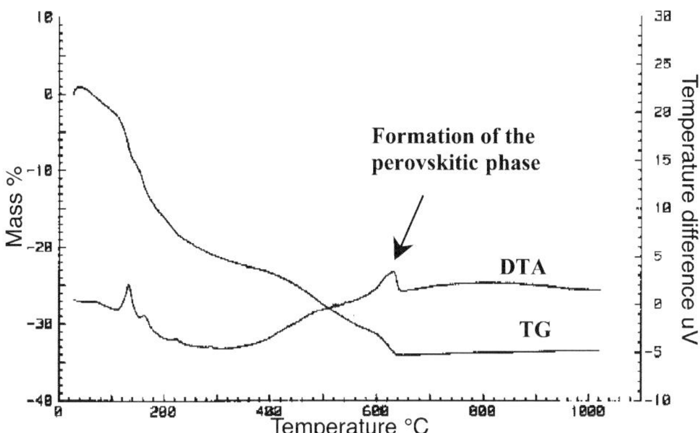

Fig. 1   DTA and TGA of the SD powder

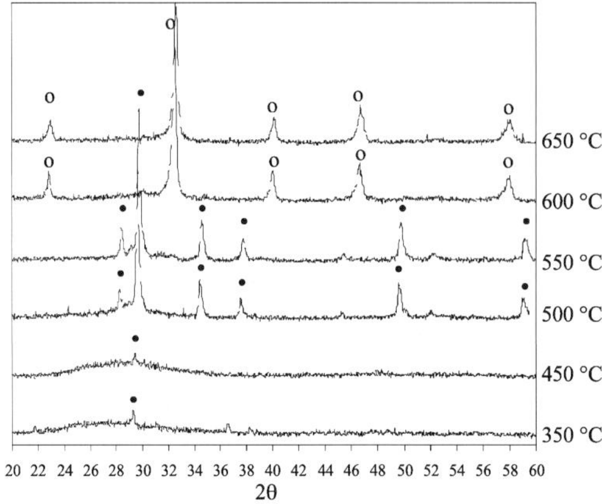

Fig. 2   XRD analysis of SD powder at different temperatures: O = perovskite, ● = pyrochlore $Bi_2Ti_2O_7$.

addition to the amorphous phase; then it changes to perovskite phase between 550 and 600°C. At 650°C the system is pure perovskitic. On the contrary, from the literature the MO powders are fully reacted only after calcination at 800°C.

The SEM images in Fig. 3 show that the process parameters and the characteristics of the salts did not result in a homogeneous volumetric precipitation: in fact the particles are essentially hollow. In Fig. 3b, it is clear that the low permeability of the salts induced explosion of

Fig. 3   SEM morphology of as spray-dried SD powder. (a) General powder; (b) detail evidencing the explosion of the particle during drying; (c) detail showing the nanoparticle morphology.

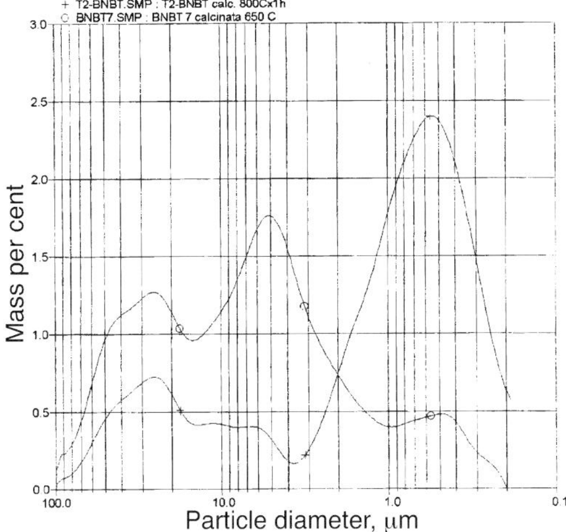

Fig. 4   Granulometric distribution of the clacined powders: O = SD, + = MM.

the particles during drying. In Fig. 3c it appears that the particles are formed by the precipitated submicrometre crystallites.

Figure 4 shows the particle size distributions of SD and MM powders calcined at 650 and 800°C, respectively, and their morphology is shown in Fig. 5. Table 5 summarises the values of specific surface and mean particle size of the powders obtained from the two techniques.

Although the two powders show the same average grain size of 0.21 μm (as calculated by the ssa values), they exhibit substantially different morphologies and degree of aggregation. The MM powder is regular in shape with a mean particle size of 0.86 μm (Fig. 5b). The SD powder shows an irregular shape (Fig. 5a) and a larger degree of aggregation with a broad particle size distribution, centred on the mean value of 6.64 μm. This kind of structure can be a result of the high reactivity of nanocrystallites forming the SD particles that can lead, at the calcination temperature, to the formation of porous aggregates of different size.

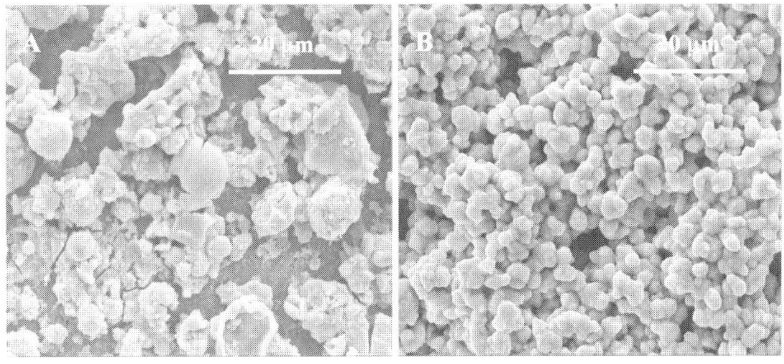

Fig. 5   SEM morphologies of the calcined powders. (a) SD; (b) MM.

Table 5 Value of specific surface area and d_{50} for SD and MM powders.

Powder	Specific surface area (m^2/g)	d_{50} (μm)
SD	4.8	6.64
MM	4.8	0.86

CONCLUSIONS

Lead-free bismuth sodium titanate–barium titanate ceramic powders were synthesised by spray drying a solution of Bi, Na nitrates, Ti isopropoxide and barium acetate or by the solid state reaction of the oxides, for comparison. The powders were compared with that obtained by the solid state reaction of the oxides. This work has confirmed the potential of spray drying as a simple synthetic process to produce chemically homogeneous powders. The samples prepared by spray drying form the pure perovskite phase at temperatures 150°C below that of conventional mixed oxide routes, however, the spraying conditions and powder processing parameters can be further improved.

REFERENCES

1. T. Takenaka, K. Maruyama and K. Sakata, '$(Na_{1/2}Bi_{1/2})TiO_3$_$BaTiO_3$ system for lead-free piezoelectric ceramics', *Jpn J. Appl. Phys.*, 1991, **30**(9B), 2236–2239.
2. T. Takenaka, A. Hozumi, T. Hata and K. Sakata, 'Mechanical properties of $(Na_{1/2}Bi_{1/2})TiO_3$ based piezoelectric ceramics', *Sil. Ind.*, 1993, **7/8**, 136–142.
3. W. D. Kingery, H. K. Bowen and D. R. Uhlman, *Introduction to Ceramics*, 2nd edn, Wiley, 1976.
4. R. W. Cahan, P. Haasen and E. J. Kramer (eds), *Materials Science and Technology: A Comprehensive Treatment*, Vol. 17A, *Processing of Ceramics part I*, Weinheim, 1991.
5. R. W. Cahan, P. Haasen and E. J. Kramer (eds), *Materials Science and Technology: A Comprehensive Treatment*, Vol. 17A, *Processing of Ceramics part II*, Weinheim, 1991.
6. G. Charlot, *Analisi Chimica Qualitativa; Equilibri in Soluzione*, 6th edn, Piccin, 1997.
7. F. A. Cotton, G. Wilkinson, C. A. Murillo and M. Bochmann, *Advanced Inorganic Chemistry*, A. Cotton and G. Wilkinson eds, 6th edn, Wiley, 1999.

Investigation of Dead Layer Thickness in SrRuO$_3$/Ba$_{0.5}$Sr$_{0.5}$TiO$_3$/Au Thin Film Capacitors

L. J. Sinnamon, R. M. Bowman and J. M. Gregg

Department of Pure and Applied Physics, Queen's University Belfast, Belfast BT7 1NN

ABSTRACT

Thin film capacitors with barium strontium titanate (BST) dielectric layers of 7.5–950 nm were fabricated by pulsed laser deposition. XRD and EDX analyses confirmed a strongly oriented BST cubic perovskite phase with the desired cation stoichiometry. Room temperature frequency dispersion ($\varepsilon_{100\ kHz}/\varepsilon_{100\ Hz}$) for all capacitors was greater than 0.75. Dielectric constant as a function of thicknesses greater than 70 nm, was fitted using the series capacitor model. The large interfacial parameter ratio d_i/ε_i of 0.40 ± 0.05 nm implied a significant dead-layer component within the capacitor structure. Modelled consideration of the dielectric behaviour for BST films, whose total thickness was below that of the dead layer, predicted anomalies in the plots of d/ε against d at the dead layer thickness. For the SRO/BST/Au system studied, no anomaly was observed. Therefore, either (i) 7.5 nm is an upper limit for the total dead layer thickness in this system, or (ii) dielectric collapse is not associated with a distinct interfacial dead layer, and is instead due to a through-film effect.

INTRODUCTION

The decrease in dielectric constant with film thickness is a well-established observation in thin film ferroelectric capacitors. This decrease can be effectively modelled by assuming the existence of very low dielectric constant 'dead layers' at the electrode/ferroelectric interfaces.[1] These interfacial dead layers act as parasitic capacitors in series with the bulk-like ferroelectric and hence the decrease in dielectric constant is said to follow the 'series capacitor model.' In this model, the effective capacitance of a thin film dielectric layer is given by

$$\frac{1}{C_{eff}} = \frac{1}{C_b} + \frac{1}{C_i} \tag{1}$$

where subscripts b and i refer to bulk and interface respectively. If the interfacial dead layer thickness d_i is independent of the total thickness d, then

$$\frac{d}{\varepsilon_{eff}} = \frac{d_b}{\varepsilon_b} + \frac{d_i}{\varepsilon_i} = \frac{d}{\varepsilon_b} + d_i\left(\frac{1}{\varepsilon_i} - \frac{1}{\varepsilon_b}\right) \tag{2}$$

It is usually assumed that either $\varepsilon_b \gg \varepsilon_i$ (Ref. 1) or $d_b \gg d_i$ (Ref. 2) giving a linear relationship between d/ε_{eff} and d with a gradient of $1/\varepsilon_b$ and y-axis intercept of d_i/ε_i.

Despite the wide acceptance of the dead layer concept, an experimentally consistent explanation for the nature of such layers is not yet evident. Various models have been proposed: low dielectric constant space charge layers;[3] oxygen depletion zones adjacent to metals with a high oxygen affinity;[4] formation of surface states;[4] local diffusion of electrode material into the ferroelectric;[5,6] lattice-mismatch-induced ion vacancy formation;[7] chemically distinct surface phase;[8] intrinsic surface polarisation effects;[9,10] depolarisation fields due to incomplete screening by the electrodes;[11,12] intrinsic suppression of polarisation at the electrode;[13] Schottky barrier, and associated depletion layer formation, as a result of band-bending at the ferroelectric/electrode interface.[14] Unfortunately none of the above is definitive nor free from contradictory evidence. Further, while direct imaging has occasionally shown distinction between bulk and interface,[15] usually high resolution transmission electron microscopy shows no evidence of interfacial layers.[14–17]

Some research suggests that a well-defined dead-layer may not be necessary for experimental agreement with the 'series capacitor' model. In this respect, the phenomenological thermodynamic approach used by Desu[18] to rationalise the existence of dead layers in fact introduces an interfacial energy per unit *area*, rather than per unit volume, such that the dead layer is not afforded a finite thickness. Despite this, the model had been successfully employed to predict the decrease in dielectric constant with thickness.[19] Basceri et al.[20] point out that although the concept of a dead layer is usually invoked to explain the observed thickness behaviour of the dielectric constant, other mechanisms could also fulfil the thermodynamic requirement. For example a long-range co-operative mechanism acting throughout the film could cause the dielectric constant to change with thickness. This is particularly pertinent given the recently observed soft-mode hardening in $SrTiO_3$ films.[21]

So, while the series capacitor model has been successfully used to fit many experimental systems it does not, in itself, conclusively demonstrate the presence of dead layers. In fact there is very little published data on the physical nature or formation of dead layers. The additional problem of the incompatibility of many published works due to differences in deposition methods and electrode materials makes it difficult to establish the behaviour of the dielectric constant over an extensive thickness range. In this paper we report the results of a comprehensive study of the effect of thickness on dielectric constant in low loss MgO/$SrRuO_3$/$Ba_{0.5}Sr_{0.5}TiO_3$/Au thin film capacitors. We show that this system generates a particularly strong parasitic capacitor component, and yet find no evidence for a distinct dead layer down to a total dielectric thickness of 7.5 nm.

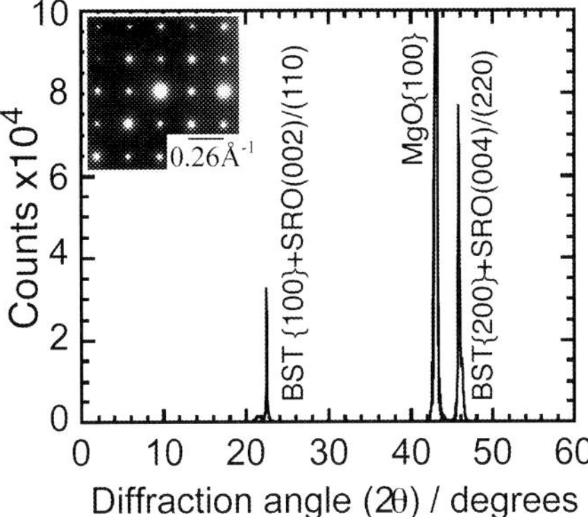

Fig. 1   Typical θ–2θ X-ray diffraction trce of SRO/BST (22 nm) structure with inplane TEM diffraction pattern of BST (inset).

METHOD

Thin film capacitors were fabricated by pulsed laser deposition (PLD) on commercial single crystal {100} MgO substrates. An energy density of ~1.5 J cm^{-2} on the target surface was supplied from a KrF excimer laser with $\lambda = 248$ nm (Lambda Physik COMP*ex* 205*i*). Growth conditions were typically a substrate temperature of 800°C and chamber pressure of 0.15 mbars O_2 for both the SrRuO$_3$ (SRO) electrode and (Ba,Sr)TiO$_3$ (BST) dielectric. All films were annealed at 650°C in 1 bar O_2 for 15 to 60 min depending on the thickness of the BST. The BST film thickness was obtained by interpolation from measurements made using an Alpha-Step 200 profilometer (Tencor Instruments) for films over 400 nm thick, and from Keissig fringes apparent on X-rays of films under 50 nm (interference between X-rays scattered from the top and bottom surfaces of the film). Thermal evaporation of Au top electrodes (radius = 0.75 mm) through a hard mask enabled dielectric measurements to be performed using an HP 4263B precision LCR meter. An Oxford Instruments cryostat and Lakeshore temperature controller facilitated dielectric measurements from 100 to 400 K.

RESULTS AND DISCUSSION

Crystallographic characterisation (Fig. 1) shows successful growth of SRO and BST perovskite phases, with strong out-of-plane (θ–2θ X-ray diffraction) and in-plane (plan-view electron diffraction inset) orientation. The out-of-plane lattice constants of BST and SRO were measured as 3.966 ± 0.004 Å and 3.93 ± 0.01 Å respectively which compare well with bulk values. Energy dispersive X-ray analysis (Jeol 733 Superprobe) showed the (Ba+Sr):Ti ratio in the BST to be close to unity (1 ± 0.03). A slight imbalance in the Sr:Ba ratio was noted (1.14 ± 0.05), giving an overall film composition of $Ba_{0.47}Sr_{0.53}TiO_3$. The effect on the dielectric constant of such a small stoichiometric change in the 'A' cation ratio is negligible.[22]

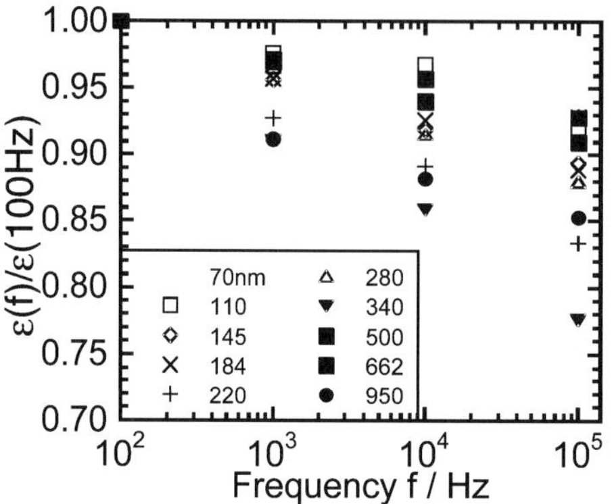

Fig. 2   Frequency dispersion of dielectric constant for capacitors of 70–950 nm.

Dispersion in dielectric constant with frequency (Fig. 2) and dielectric loss values (Fig. 3a) were monitored, and found comparable to literature.[17,23] Dielectric constant as a function of BST film thickness is shown in Fig. 3a, taken at 400 K so that inherent thickness effects could be distinguished from Curie anomaly suppression (for bulk $Ba_{0.5}Sr_{0.5}TiO_3$ T_c = 248 K (Ref. 24)). Very thin capacitors showed a flat temperature dependence of dielectric constant and tan δ. (Except for the occasional result with enhanced ionic conduction above 350 K. For these capacitors lower temperature values were used.) The absolute magnitude of the dielectric constants are somewhat lower than reported for other BST thin film systems.[25,26] This suppression appears to be caused by the BST/Au interface, since measurements made on the *same* BST sample using both Au and Pt top electrodes revealed up to a threefold increase in the dielectric constant associated with Pt compared with that measured using Au top electrodes. For our capacitor structures, with $d > 70$ nm, values extracted from Fig. 3b using the series capacitor model were ε_b = 1000 ± 200 and d_i/ε_i = 0.40 ± 0.05 nm. The solid line is the

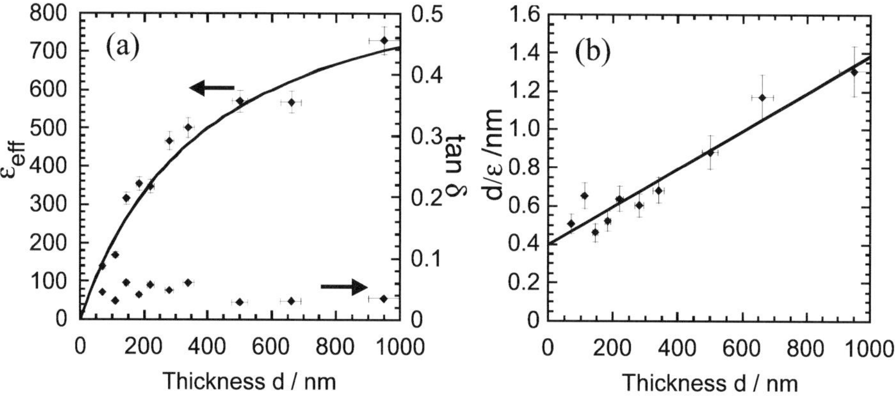

Fig. 3   (a) Dielectric constant and tan δ as a function of BST thickness from 70–950 nm measured at 400 K. The solid line shows the fit generated using the model parameters extracted from part b. (b) d/ε as a function of BST thickness d from 70–950 nm. The solid line is the linear fit to the data (R^2=0.93).

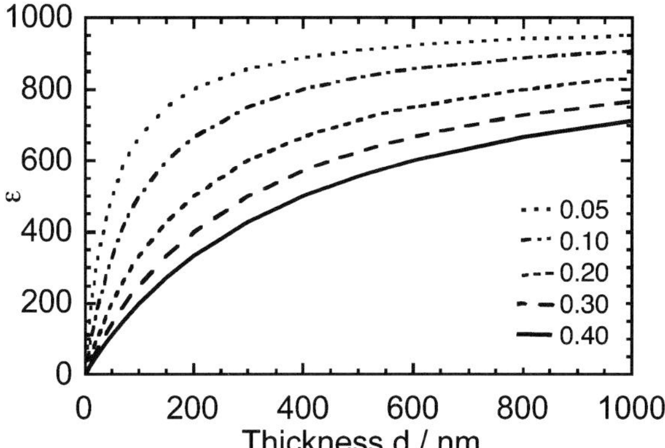

Fig. 4   Effect of d_i/ε_i value on decrease of ε with thickness.

fit generated from these values showing that the system follows the model and thus acts as if dead layers are present. The value of the d_i/ε_i ratio controls the decrease in ε with thickness as shown by Fig. 4. This d_i/ε_i ratio is relatively large (compared with 0.05 nm for Pt/BST/Pt (Ref. 27), 0.13 nm for Pt/PZT/Al and 0.08 nm for Ag/PZT/Ag (Ref. 28), 0.12 nm for Pd/BST/Au/Ti (Ref. 29) and 0.887 nm for Al/BST/Al (Ref. 30)) and the effect of the parasitic capacitor component is therefore apparent even in relatively thick films. However, the fact that d_i and ε_i are inseparable in the conventional series capacitor model, means that neither the dead layer thickness nor the dielectric constant is well defined.

It was hypothesised that if a physically distinct dead layer exists, then when the total thickness of the dielectric is less than the thickness of the dead layer, some deviation from the conventional series capacitor model would be expected. The model defines the dead layer as having a constant ε_i, thus when $d > d_i$, ε_{eff} is given by the combination of the layers as prescribed by equation (2), but when $d \leq d_i$, ε_{eff} becomes constant (ε_i). This case is shown schematically as (i) in Fig. 5. This scenario which produces a step function in dielectric

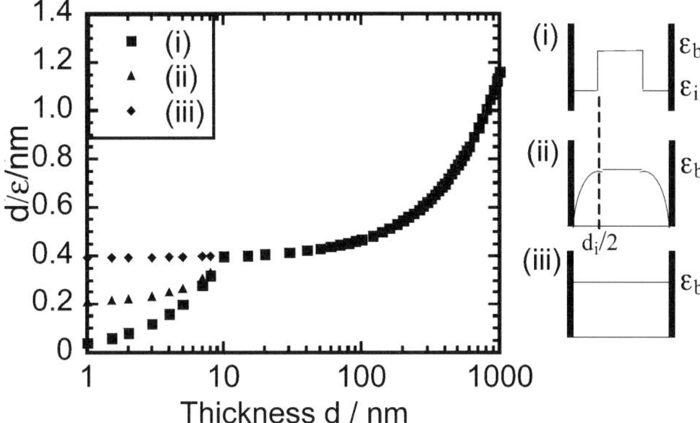

Fig. 5   Simulations of d/ε behaviour using the series capacitor model for the cases when (i) ε_i is constant; (ii) ε_i decreases parabolically, (iii) there is no distinct dead layer. For (i) and (ii), the total thickness of the dead layer d_i was set at 10 nm.

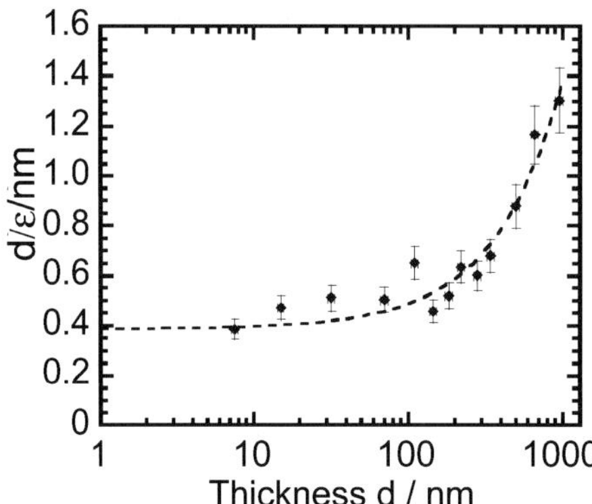

Fig. 6 d/ε as a function of BST thickness down to 7.5 nm. The function expected for a dead layer of zero thickness is shown by the dashed line using the parameters from Fig. 3b.

constant through the film is physically unlikely so we have considered a more continuous fall-off function (parabolic reduction in ε_i within the dead layer). In this case when $d \leq d_i$, $\varepsilon_{\mathit{eff}}$ is given by

$$\varepsilon_{\mathit{eff}} = -\frac{\varepsilon_b d}{d_i \left[\left(\dfrac{d}{d_i}\right) - 2\right]} \tag{3}$$

As can be seen in Fig. 5, in both of these cases a sharp anomaly is exhibited in the d/ε against $log(d)$ function at $d = d_i$. The case of the dead layer being attributed no finite thickness is added to (iii) in Fig. 5 to demonstrate that the anomalies generated in (i) and (ii) do not occur, and hence that an observed anomaly could be confidently interpreted as yielding a value for d_i.

Capacitors with BST thicknesses less than 70 nm were grown to investigate the presence of an anomaly and hence demonstrate the dead layer thickness in the SRO/BST/Au system. Although the tan δ values increased to 0.6 at 10 kHz for the 7.5 nm film, the frequency dispersion was unchanged. As can be seen in Fig. 6, the experimentally determined d/ε against $log(d)$ plot of all results from 7.5–950 nm shows that the dielectric behaviour for films below 70 nm obeys the series capacitor model with the same parameters as those extracted from the series above 70 nm. The expected curve for the case of no dead layer is shown by the dashed line which must have a y-axis intercept of (0.40 ± 0.05) nm, according to the previous analysis. As can be seen, no anomaly in the experimental results is observed for BST thicknesses down to 7.5 nm.

CONCLUSIONS

The results in Fig. 6 show no evidence for the presence of dead layers down to a thickness of 7.5 nm, despite the expectation that the SRO/BST/Au system should possess a substantial parasitic component. Therefore either:

(i) dielectric collapse in thin film ferroelectrics is not associated with a distinct interfacial dead layer, and is instead due to a through-film effect, or

(ii) 7.5 nm represents an upper limit on the total dead layer thickness for the SRO/BST/Au system, with an associated maximum average ε_i of ~19.

Further work is under way using much smaller electrode areas and much thinner BST films to experimentally verify or otherwise the existence of dead layers with finite thickness.

ACKNOWLEDGEMENTS

The authors gratefully acknowledge financial support from DFHETE, the EPSRC, NICAM and The Royal Society, as well as the collaboration of AVX/Kyocera.

REFERENCES

1. K. Amanuma, T. Mori, T. Hase, T. Sakuma, A. Ochi and Y. Miyasaka, *Jpn J. Appl. Phys.*, 1993, **32**, 4150.
2. H. Li, W. Si, A. D. West and X. X. Xi, *Appl. Phys. Lett.*, 1998, **73**, 464.
3. V. G. Bhide, R. T. Gonhalekar and S. N. Shringi, *J. Appl. Phys.*, 1965, **36**, 3825.
4. G. Teowee, C. D. Baertlein, E. A. Kneer, J. M. Boulton and D. R. Uhlmann, *Integrated Ferroelectrics*, 1995, **7**, 149.
5. I. Stolichnov, A. Tagantsev, N. Setter, J. S. Cross and M. Tsukada, *Appl. Phys. Lett.*, 1999, **75**, 1790.
6. D. Choi, B. Kim, S. Son, S. Oh and K. Park, *J. Appl. Phys.*, 1999, **86**, 3347.
7. M. Izuha, K. Abe and N. Fukushima, *Jpn J. Appl. Phys.*, 1997, **36**, 5866.
8. V. Craciun and R. K. Singh, *Appl. Phys. Lett.*, 2000, **76**, 1932.
9. C. Zhou and D. M. Newns, *J. Appl. Phys.*, 1997, **82**, 3081.
10. K. Natori, D. Otani and N. Sano, *Appl. Phys. Lett.*, 1998, **73**, 632.
11. P. Wurfel and I. P. Batra, *Phys. Rev. B*, 1973, **8**, 5126.
12. Y. G. Wang, W. L. Zhong and P. L. Zhang, *Phys. Rev. B*, 1995, **51**, 5311.
13. O. G. Vendik, S. P. Zubko and L. T. Ter-Martirosayn, *Appl. Phys. Lett.*, 1998, **73**, 37.
14. C. S. Hwang, B. T. Lee, C. S. Kang, K. H. Lee, H. Cho, H. Hideki, W. D. Kim, S. I. Lee and M. Y. Lee, *J. Appl. Phys.*, 1999, **85**, 287.
15. S. Paek, J. Won, K. Lee, J. Choi and C. Park, *Jpn J. Appl. Phys.*, 1996, **35**, 5757.
16. Y. Sakashita, H. Segawa, K. Tominaga and M. Okada, *J. Appl. Phys.*, 1993, **73**, 7857.
17. W. Jo, D. C. Kim, H. M. Lee and K. Y. Kim, *J. Korean Phys. Soc.*, 1999, **34**, 61.
18. S. B. Desu, *Mater. Res. Soc. Symp. Proc.*, 1999, **541**, 457.
19. K. Abe and S. Komatsu, *Jpn. J. Appl. Phys.*, 1993, **32**(2), L1157.

20. C. Basceri, S. K. Streiffer, A. I. Kingon and R. Waser, *J. Appl. Phys.*, 1997, **82**, 2497.

21. A. A. Sirenko, C. Bernhard, A. Golnik, A. M. Clark, J. Hao, W. Si and X. X. Xi, *Nature*, 2000, **404**, 373.

22. T. Nakamura, Y. Yamanaka, A. Morimoto and T. Shimizu, *Jpn J. Appl. Phys.*, 1995, **34**, 5150.

23. B. Nagaraj, T. Sawhney, S. Perusse, S. Aggarwal, R. Ramesh, V. S. Kaushik, S. Zafar, R. E. Jones, J. H. Lee, V. Balu and J. Lee, *Appl. Phys. Lett.*, 1999, **74**, 3194.

24. K. H. Hellwege (ed.), *Landolt-Börnstein Numerical Data and Functional Relationships in Science and Technology*, New Series Group III: Crystal and Solid State Physics, Vol. 16a, Springer-Verlag, 1981.

25. W. Lee, I. Park, G. Jang, and H. Kim, *Jpn J. Appl. Phys.*, 1995, **34**, 196.

26. M. Izuha, K. Abe, M. Koike, S. Takeno and N. Fukushima, *Appl. Phys. Lett.*, 1997, **70**, 1405.

27. W. Lee, H. Kim and S. Yoon, *J. Appl. Phys.*, 1996, **80**, 5891.

28. J. J. Lee and S. B. Desu, *Ferroelectric Lett.*, 1995, **20**, 27.

29. S. Yamamichi, H. Yabuta, T. Sakuma and Y. Miyasaka, *Appl. Phys. Lett.*, 1994, **64**, 1644.

30. N. Ichnose and T. Ogiwara, *Jpn J. Appl. Phys.*, 1995, **34**, 5198.

Depth-Tolerant Transducer Technology

G. Smith

*Thales Underwater Systems, Redfields Industrial Park, Church Crookham, Fleet, Hampshire
GU52 0RD*

ABSTRACT

We have conducted a design study to identify and prove sonar projector technology capable of meeting strict criteria, including tolerance of considerable depth without significant performance change. Various candidate transducer technologies were examined. Of these, the air-backed bender was well-known to be capable of providing good acoustic performance at low cost, in a body of low mass and small dimensions. It was, however, traditionally difficult for a small-diameter air-backed bender projector simultaneously to provide low resonance frequency, and adequate bandwidth, with operation at substantial depth. The study successfully culminated in the build and test of a reasonably-priced, high-power, air-backed, omnidirectional bender projector, resonant at 2260 Hz in water, with a −3 dB bandwidth of 225 Hz, and a measured maximum efficiency of 80%. It was proven satisfactory to 500 m, with no significant performance deterioration. As a bonus, it also functioned as a high-sensitivity hydrophone, with substantial capacitance. The design was capable of further development and cost reduction. An array of four such projectors could produce a source level of 210 dB re 1 µPa at 1 m. This array had a packed length of 144 mm, a mass of 4.8 kg (excluding cable) and was suitable for deployment from an A-size canister.

INTRODUCTION

Thales Underwater Systems (TUS) had a requirement for a projector resonant at ≈2200 Hz in water, with a −3 dB bandwidth of ≈200 Hz. This device had to operate, for several hours, at considerable depth in the ocean, given a maximum outside diameter of 111 mm, and low mass. Electroacoustic efficiency had to be as high as possible because limited battery power was available. An array of these projectors had to provide a maximum source level of ≈210 dB re 1 µPa at 1 m, in long pulses. The packed length of the array elements had to be as small as possible. Several candidate technologies were identified, and are discussed below. Two transducer technologies were preferred. Of these, the air-backed bender concept was addressed, and is the main subject of the present paper.

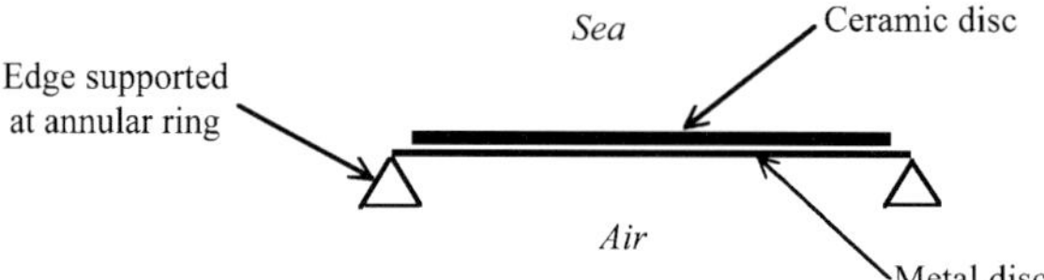

Fig. 1   Single bilaminar air-backed bender.

PROJECTOR TYPES

Benders

Bender transducers are based upon at least one thin active ceramic element. They may be air-backed or liquid-backed.

Air-backed bender

Air-backed benders are characterised by relatively narrow bandwidth, and their performance is depth-limited. The cheapest would be the 'single bilaminar' type shown in Fig. 1. The combination of ceramic and metal discs is here termed the 'bender element'. The air is contained, in a pocket, behind the bender element, by an acoustically inert blanking plate (not shown). For deep-ocean service, the ceramic disc must be located on the outside, so hydrostatic depth pressure will only subject it to compressive stress (tensile stress would destroy the ceramic); the inability to locate an additional ceramic disc on the inside of the metal disc limits the electrical power handling capability. Depth-performance could be improved by pressurising the contained air (or other gas) in step with depth. This was considered too complicated and expensive to implement. A 'double bilaminar air-backed bender' is similar to its 'single' cousin, except that two similar devices are mounted back-to-back, and the requisite air pocket is contained between them. Clearly, this has twice the electrical power handling capability of the equivalent 'single' device, at some additional cost.

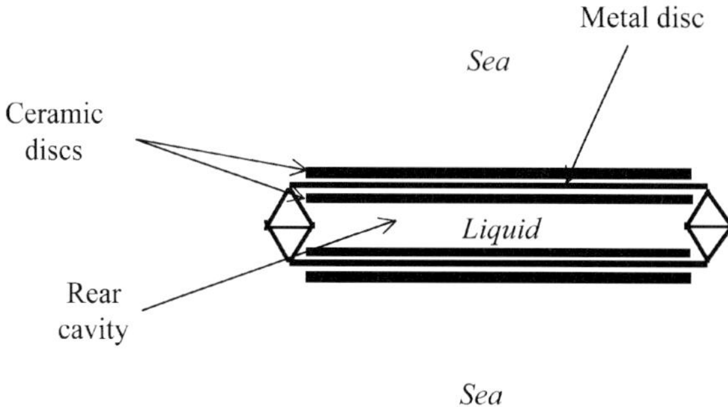

Fig. 2   Double trilaminar liquid-backed bender.

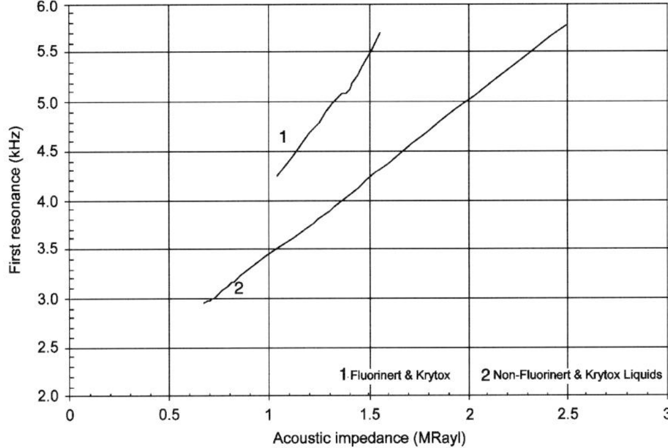

Fig. 3   Rear cavity fill liquid pc versus resonance frequency (length of rear cavity 100 mm and diameter 100 mm).

Liquid-backed bender

Liquid-backed benders exist in the same classes as air-backed benders. They are depth-independent, so an additional ceramic disc may be located upon the inside of the bender element (because hydrostatic depth pressure cannot subject it to tensile stress) doubling the power handling capability (*see* Fig. 2). The liquid-filled cavity should be provided with a breathing hole to permit equalisation with ambient pressure. While the main resonance of an air-backed bender is set by the bender element, in a liquid-backed bender it is set by the rear cavity. If the bender element's resonance coincides with that of the cavity, Q_m will be high. The further apart the resonances, the lower Q_m will be. The resonance frequency, and bandwidth, of a liquid-backed bender can, therefore, be tailored by modifying the length of the rear cavity, and by changing its liquid fill. Using Woollett's equations,[1] the resonance frequency of the rear cavity bears the interesting relationships with the acoustic impedance of the fill liquid shown in Fig. 3 and Table 1. No such linear relationship of resonance frequency with density, or sound speed alone, was found.

Table 1   Data used in Fig. 3.

Liquid	Acoustic impedance (MRayl)	First resonance (kHz)	Liquid	Acoustic impedance (MRayl)	First resonance (kHz)
Glycerol	2.495	5.8	Fluorinert FC-5311	1.552	5.7
Transformer oil	1.36	3.8	Krytox K-1525	1.381	5.1
F111/50 silicone oil	9.803	3.2	Krytox K-1514	1.359	5.1
MS200 (2.0) silicone oil	8.102	3.0	Fluorinert FC-70	1.333	5.05
MS-200 (0.65) silicone oil)	0.6636	2.96	Fluorinert FC-75	1.029	4.25

Fig. 4   Double trilamminar bender projector with liquid-filled cavities front and rear.

Changing the cavity fill-liquid, therefore, did not offer the possibility of achieving 2200 Hz. The longer the cavity, the lower the resonance frequency; but it was found that the cavity would have had to be ≈300 mm long to attain 2200 Hz (using silicone oil MS-200 (0.65)), which was too great to meet the design requirement. A liquid-backed bender would, therefore, have been too heavy, too long and too expensive to meet the design requirement. The bandwidths of both air- and liquid-backed benders can also be tailored by the provision of a liquid-filled cavity on the front of each bender element (*see* Fig. 4). This can also provide enhanced acoustic coupling with the ocean.

Cylindrical bender
The cylindrical bender's active elements are coaxially-oriented bender bars located around the perimeter of the cylinder[2] (*see* Fig. 5). Each bender bar element comprises two beams, one of active ceramic, the other of steel to withstand depth pressure, the former located on

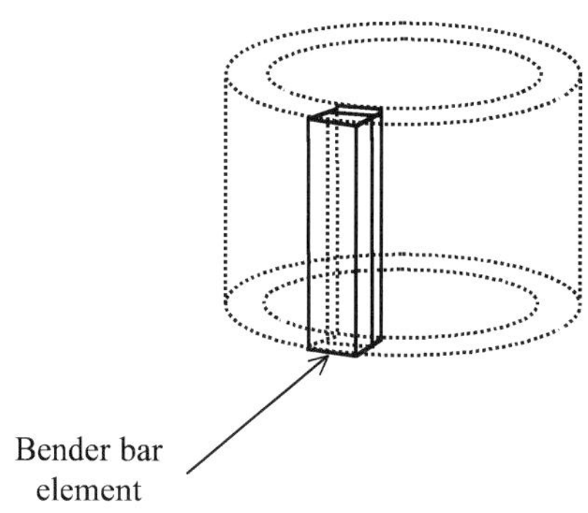

Fig. 5   Cylindrical bender.

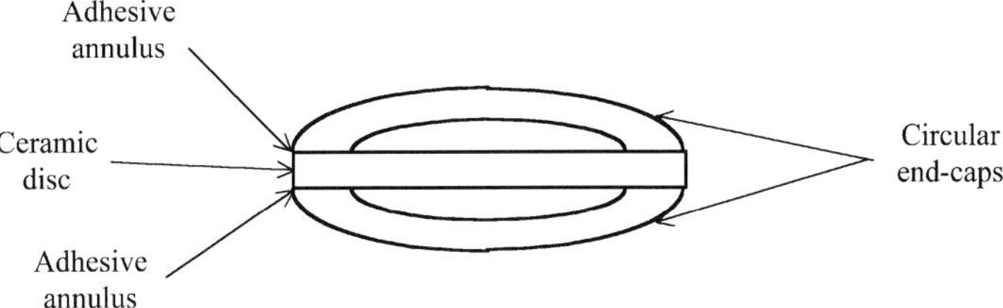

Fig. 6 Clam shell concept.

the outside to prevent its exposure to depth-induced tensile stress. The resonance frequency is not set by the device's diameter, so an example could have been designed to fit into an A-size cylindrical canister, in which case the acoustic field would have been omnidirectional. Were the projector air-filled, however, its depth-performance and bandwidth would have been limited. The mass was moderate, and the transmit sensitivity adequate. Efficiency was relatively low. The concept was rejected as expensive, and high-risk for the limited development time available.

Clam shell

We termed the transducer type shown in Fig. 6 'clam shell'.[3] Multilayer projectors (e.g. piston transducers) exhibit high force, but small displacement; whereas cantilever bimorph projectors (e.g. benders) exhibit large displacement, but small force. Used as projectors, clam shells exhibited exceptionally high displacement and force. Efficiencies greater than 95% were claimed, with remarkably high effective coupling coefficients. The design was potentially cheap, as the end-caps could be simultaneously cut and shaped by die punching.

TUS AU (Australia) had built and tested similar devices for use as hydrophones. In the present study, a design exercise was conducted to estimate the concept's suitability for the present application as a high-drive projector, using Navy Type III ceramic[4] with a central hole to reduce the main resonance to $\approx$2200 Hz, and providing reinforcement to preserve structural integrity under high drive. It was calculated that the maximum output acoustic power would be $\approx$1140 Watt; had this been achieved, attention would have had to be paid to the possibilities of overheating, and of cavitation in shallow depth operation. Assuming zero directivity, the maximum source level would have been $\approx$201 dB re 1 μPa at 1 m, and the maximum transmit sensitivity $\approx$11.5 dB re 1Pa/V at 1 m. This established the feasibility of constructing an evaluation device, though the risk was considered high for the limited development time available.

Liquid-filled tube

Conventional design
An active ceramic tube has an important resonance mode, here called the 'circumferential' mode, whose value is controlled by the tube's diameter. The requirement to fit the projector

inside an A-size canister obviated the possibility of using the circumferential mode in the required frequency range. Were the tube liquid-filled, it would have another important resonance môde, the 'cavity' mode, whose resonance frequency is controlled by the length of the tube, whose maximum efficiency is, however, only $\approx 40\%$. The cheapest method of achieving liquid-fill is to permit free-flooding with sea water. Were a fill-liquid of lower acoustic wavespeed than water selected, the length of the tube could be reduced, saving space. It is possible to fill tube projectors with a polyurethane selected for low mechanical tan δ, and for reasonable acoustic impedance match with water. Such a device is, however, characterised by relatively high loss, which would have reduced the efficiency still further. In tube transducers (unlike most other types), conductance Q is much greater than transmit sensitivity Q (especially in the cavity mode). Conductance Q can be reduced by means of a suitable matching network; however, the cavity mode resonance frequency will shift significantly with changing water temperature, detuning the network. The amplitude of the conductance peak will also vary considerably with the degree of wetting of the projector (probably not a problem, under high drive).

The ceramic tube can be a single piece, radially polarised, or tangentially polarised via painted silver electrodes. Alternatively, it can be assembled from segments ('barrel stave' construction), in which case the electrodes are located at the joints between the segments, and polarisation is tangential. If high electrical drive is to be used, it is customary to apply tangential prestressing by means of GFRP, typically to ≈ 20 MPa. Radial polarisation implies driving the ceramic tube in the 1-3 mode, while tangential polarisation implies use of the more efficient 3-3 mode. The striped tube construction offers a compromise between performance and cost (although a monolithic ceramic tube can only be made up to a diameter of about 150 mm). The acoustic performance details, of various designs of free-flooded ceramic tube projector, are given in Table 2. From Table 2, the maximum source level of a

Table 2   Measured acoustic performance of free-flooded ceramic tube projectors. All were made of Navy Type III ceramic. All were prestressed to 20 MPa nominal tangential stress. The stripes were all 2 mm wide. R is radially polarised; S is striped; B is barrel stave; S_{max} is maimum transmit sensitivity; SL_{max} is maximum source level.

Type	Diameter (mm)	Length (mm)	Thickness (mm)	No. of segments	S_{max} (dB re 1Pa/V @ 1m)	SL_{max} (dB)
R	73.0	38.0	3.0	1	13.8	Ref.
S	73.0	38.0	3.0	20	8.2	+3.9
R	63.4	31.3	5.0	1	13.4	Ref.
B	66.4	34.5	4.4	18	11.9	+4.9
R	33.5	16.8	2.3	1	13.2	Ref
S	34.7	17.6	2.3	10	6.8	+3.8
B	34.8	18.1	2.3	12	8.3	+4.8

striped tube will only be about 1 dB below that of a barrel stave tube, although a striped tube will be $\approx 4^{1}/_{2}$ times cheaper to purchase than the equivalent barrel staved tube. Radially-polarised tubes will be cheaper than either, with reduced performance. It was assessed that to produce a cavity mode resonance of ≈ 2200 Hz would require a ceramic tube of length ≈ 118 mm, which was too long to meet the design requirement, and would have been too expensive.

Variations upon liquid-filled tube

The cavity mode resonance frequency of a liquid-filled tube transducer can be lowered by lengthening the tube by means of acoustically inert extension(s).[5] These would have had to be nested with the ceramic tube, to reduce the undeployed length. The presence of electrodes and wiring would have had to be taken into account; and it would have been essential to ensure good acoustic and mechanical connection between the ceramic tube and the extensions, or they would have functioned as separate radiators. Given the time available, these difficulties were considered to raise the risk to an unacceptable level. A variant upon the concept is to bore circumferential arrays of holes in the tube extensions to increase the radiation between the cavity resonances, thus broadening the frequency band.[6] We had built and tested a magnetostrictive tube projector, with aluminium alloy tube extensions, and found it satisfactory; but it was too large for the present application. A design of this type was commercially available; but was also too large.

PVdF ring

We had built and tested a ring projector constructed of PVdF film wound upon a cylindrical Perspex former. This resonated in the frequency range required for the present application; however, the transmit sensitivity was too low to meet the design requirement.

Slotted tube

Also called 'split ring' transducers, this concept[7] has the general form shown in Fig. 7, which represents an active ceramic ring or tube which may, or may not, be reinforced by an external tube of similar form, typically of aluminium alloy or GFRP. The device is air-filled, so the depth limit is set by the closure of the slot. Extra-mural designs, which would have fitted inside the A-size canister, had low mass, and maximum source level up to 197 dB re 1 μPa at 1 m. Q_m was ≥ 5, and the efficiency was $\leq 90\%$. This option was thoroughly investigated; but was rejected because the projectors were longer than the air-backed bender projector which we developed, and because they were expensive.

Flextensional

Most classes of flextensional projector are expensive, though large-scale production might reduce cost to an acceptable level. They are usually characterised by high Q_m and poor power factor, while a feature of the overall design requirement was the limited battery power

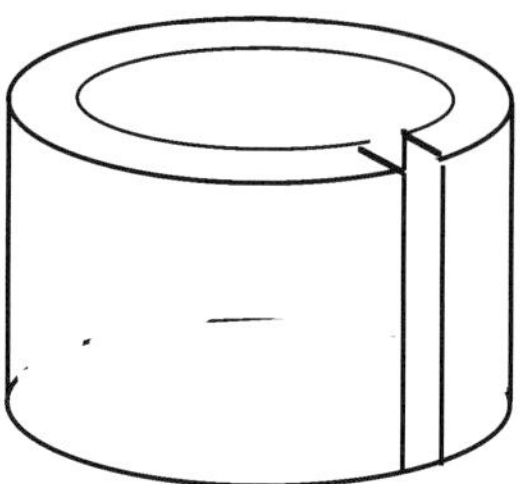

Fig. 7   Slotted tube.

available; a suitable matching network would improve the situation, but increase the complication and cost. The most promising available device was a Class I 'Diabolo' originally developed by TUS FR (France), and produced by TUS AU (Australia). This is also called an 'inverse barrel stave' device, and has the form shown in Fig. 8. The concave segments are sprung to apply compressive stress to the ceramic stack. Classes of flextensional projector, prestressed by convex segments, suffer progressively reduced stack compression, with increasing depth, leading to their eventual destruction. A Class I device has the advantage that it is unlikely to be destroyed by depth pressure; but that the stack will merely 'lock up', at about 300 m, without an internal pressure-compensation system.

At 10 m depth, the Diabolo was resonant at 1550 Hz. The resonance varied by 300 Hz between 4 and 250 m (a feature common to flextensionals). The bandwidth was 300–400 Hz. The maximum drive was ≥ 2.1 kV$_{rms}$ (2.6 kV$_{rms}$ for a limited lifetime), a single device yielding a maximum source level of ≈ 193 dB re 1 µPa at 1 m, in an omnidirectional acoustic field. The efficiency was 45–50% (greater, in an array), and the effective coupling coefficient was 0.25. Diabolo was 163 mm long (excluding cable and connector). It had a diameter of 55 mm, permitting three examples to be packed in parallel, in an A-size canister, prior to deployment. Its mass was ≈ 1 kg. Its price was reduced to the point at which it became an attractive option.

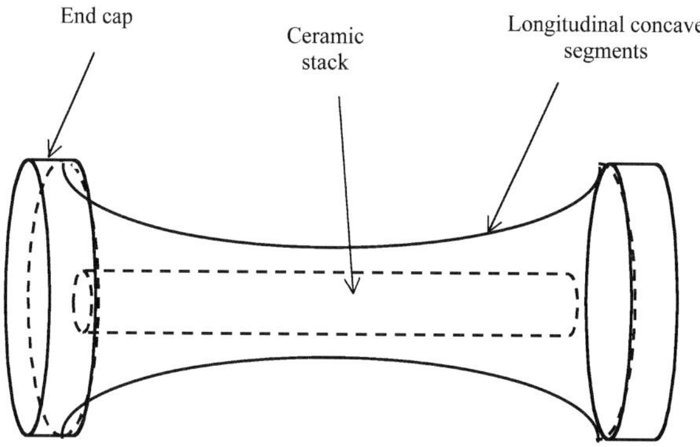

Fig. 8   Diabolo.

Electrodynamic

Also called a 'moving coil' transducer, a single device can provide useful source level from a few tens of Hertz to low kiloHertz, and it can be tuned. Although this type of projector was commercially available, it was too expensive, and too large, for the application.

DRIVE MATERIALS

Given the implied requirement for high performance from a small-volume device, consideration was given to novel active materials such as relaxors and phase-change materials; also, single crystals had started to become commercially available, and magnetostrictives had been available for some time. It was considered that we should not proceed directly to construction and testing of high-power projectors, based upon novel ceramics, because:

1. some had not been sufficiently well characterised or developed (e.g. relaxors)
2. some were not expected to function with sufficient efficiency across the full operational temperature range of -2 to $+35°C$ (e.g. transition temperature limits of relaxors)
3. some had frequency response limits (e.g. phase-change materials were unlikely to yield high strain output above about 1000 Hz)
4. some were excessively time-consuming to produce in production quantities (single crystals)
5. dc biasing would be required to obtain the full potential of some novel materials (an unwelcome complication)
6. availability of some novel materials was limited (especially single crystals)
7. some were expensive (single crystals and, to a lesser extent, Terfenol-D)
8. Navy Type III ceramic was, therefore, preferred to meet the requirement for long pulses at high drive, at the development model stage.

SUMMARY OF CANDIDATE PROJECTOR TECHNOLOGIES

The candidate projector technologies are summarised in Table 3.

Preferred solution

Discussion of preferred candidate solutions
The preferred candidate design concepts were Class I flextensional (Diabolo) and air-backed bender. The former was pursued by TUS AU and the latter was the subject of the present study.

Implementation of air-backed bender concept[8]
As it stood, the air-backed bender design concept could not meet the design requirement. It was traditionally difficult to achieve low resonance frequency simultaneously with relatively

Table 3   Technical summary of projector concepts examined (1997/1998). Source level measured at resonance. Efficiency expressed as useful acoustic power into the water versus electrical drive power.

Projector type	Resonance frequency (Hz)	Q_m (approx)	Number of resonances	Source Level (dB re 1µPa @ 1m)	Max efficiency	Max depth (m)	Weight in water	Dimensions	Cost	Risk
Air-backed bender	>2000 - 5000	10 to 20 / adjustable	Single	200	85%	200 m (conventional)	Low	OK	Low	Low
Liquid-backed bender	>2000 - 5000	Adjustable over wide range	Single / multi	<200	65%	OK	High	Too long	Medium	
Cylindrical bender	OK	10 to 20	Single	High	65%	Unknown		Longer than air-backed bender	Medium / high	High
Clam shell	<2000 - 5000	Unknown	Single	203	60%?	OK				
Liquid-filled tube	>2000 - 5000	5 to 10	Single	207 to 211	40%	OK	Medium			
Ditto with extension(s)	500 - 5000	5 to 10	Single	As above	As above	OK	OK			
Slotted tube	150 - 7000	1 to ≥20	Single / multi	177 to 197	90%	120 m	Low	Longer than air-backed bender	Medium	Low
Flextensional Classes I to III	1300 to 1700	5	Single / multi	191	45% (65% in array of 6)	200 m to 300 m	Low	OK	Medium	Low
Flex Classes IV & VII	500 to 4000	3	Single/ multi	190 to 195	60%	400 m (pressure compensation)	Low (eg: 2 kg)	Wrong shape	High	Low
Flex Classes V & VI	155 to 2700	4 to 11	Single/ multi	205 to 217	65% to 90%	250 m (pressure compensation)	Low	Too large	High	Low
Moving Coil	30 - 10000	Broadband	Tuneable	165 to 190	Low	300 m (pressure compensation)	High (eg: 82 kg)	Too large	High	Low

small diameter and good depth performance.[1] Good depth performance meant a thick bender element, which also implied high resonance frequency. Limited diameter also implied a lower limit to the resonance frequency. The depth limit was set by the onset of depolarisation of the ceramic discs, which is stress-induced. Finite element analysis was used to design the device such that this stress was minimised. High-drive operation at shallow depth bore the risk of blasting the bender element from the device, so it was essential to have good adhesion between the backing plate and the annular support ring. This could be achieved by welding, hard or soft soldering, or chemical adhesive. Welding or soldering might damage a thin backing plate, and is more expensive than adhesive. Experience indicated that an adhesive bond would provide adequate bond strength, given sufficient bond area, so this option was pursued.

Since time was available only for one design-and-build iteration, a belt-and-braces approach was pursued to maximise the chance of demonstrating the candidate technology's feasibility for meeting the design requirement. This incurred cost, which could be shed in a production model. A consequence was that the prototype backing plates were made of maraging steel, whose characteristic high strain-to-failure was expected to be of value at the thin annular hinge. Subsequent to machining, this steel required special heat treatment, which was accomplished without distortion.

Detailed finite element analysis confirmed that the novel concept was viable, and predicted acceptable performance, including a depth limit of 500 m, for recoverable performance. Referring to Figs 9–12, acoustic testing, in a large body of water, confirmed the main performance predictions, to close tolerances. Transmit efficiency was measured as acoustic power into the water versus input electrical power. Near resonance, the device had a directivity index of ≈ 1 dB. Capacitance at 1000 Hz was 17 nF, and tan $\delta \approx 0.006$. The effective coupling coefficient was ≈ 0.2.

The depth limit was proved by measuring the receive sensitivity, in a pressure vessel, by means of white noise techniques. At a pressure equivalent to 500 m, the receive sensitivity

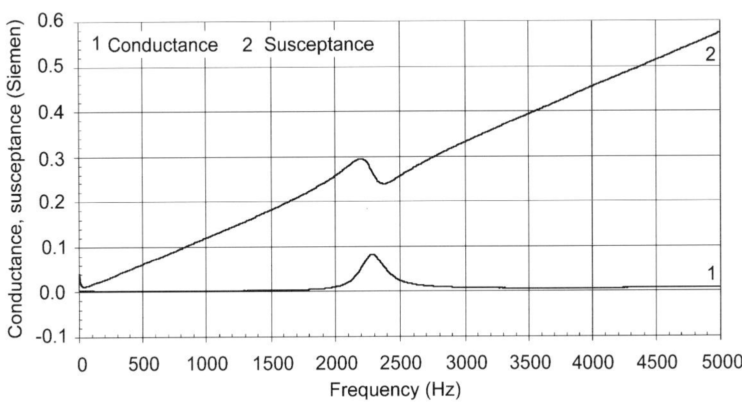

Fig. 9   In-water admittance characteristics of single depth-tolerant bender projector.

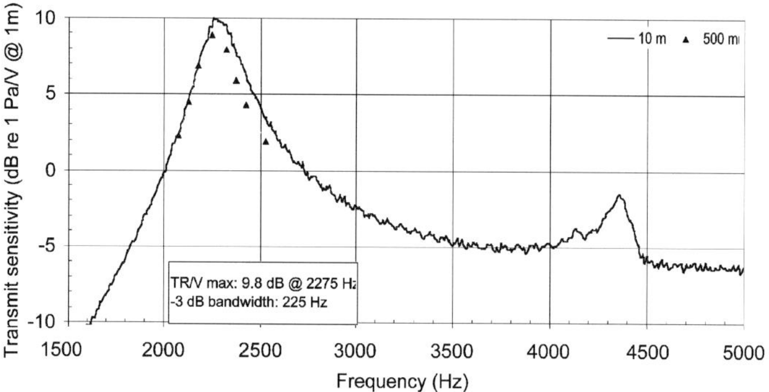

Fig. 10   Transmit sensitivity of single depth-tolerant bender projector.

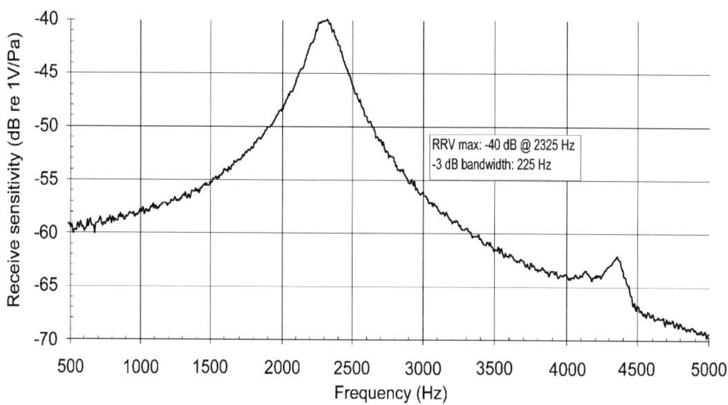

Fig. 11   Receive sensitivity of single depth-tolerant bender projector.

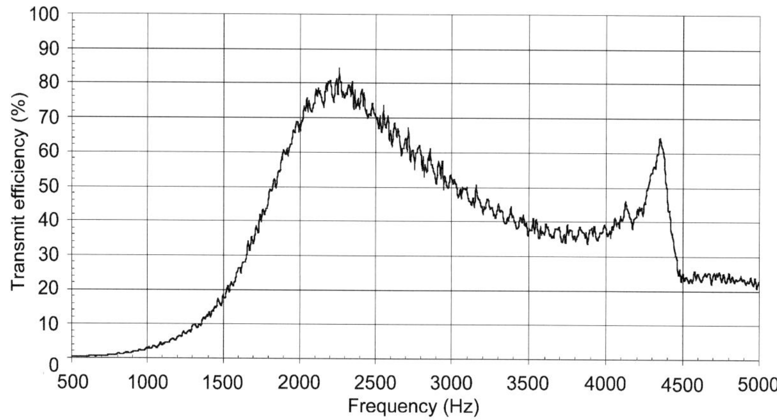

Fig. 12   Transmit efficiency of single depth-tolerant bender projector.

Fig. 13   Depth-tolerant bender projector.

had reduced by 1 dB (equivalent to a similar fall in transmit sensitivity), and the resonance frequency had fallen by 50 Hz. The value measured at ambient was recovered upon pressure relaxation, even following several pressurisation cycles. Previous experience, with similar devices, indicated that a 1 dB sensitivity change signposted the maximum depth at which the depth-tolerant bender's performance would be recoverable. Figure 13 is a photograph of a prototype depth-tolerant bender projector.

CONCLUSIONS AND SUBSEQUENT WORK

The work described herein proved concept feasibility, as required.

REFERENCES

1.   R. S. Woollett, 'Theory of the piezoelectric flexural disk transducer with applications to underwater sound', United States Underwater Sound Laboratories Research Report 490, S-F001 03 04-1, 5th December 1960.
2.   'Cylindrical Bender-Type Vibration Transducer', UK Patent GB2173670B.
3.   A. Dogan, K. Uchino and R. E. Newnham, 'Composite piezoelectric transducer with truncated conical endcaps 'Cymbal'', *IEEE Trans. Ultrasonics, Ferroelectrics Frequency Control*, 1997, **44**(3), 597–605.
4.   'Piezoelectric ceramic material and measurements guidelines for sonar transducers', Military Standard (MIL-STD) -1376.
5.   V. E. Glazanov and A.V. Mikhailov, 'Method to construct a low-frequency emitter by using

water-filled tubes', *Acoustical Phys.*, 1996, **42**(5), 537–542.
6.   B. L. Flanning, G. W. McMahon and D. D. Prentiss, *A Vented Resonant-Pipe Projector,* Canadian Ministry of Supply and Services, 1987.
7.   'Electro-Acoustic Transducer', USA Patent 5229978, 20th July 1993.
8.   European Patent 1041537A2.

Comparison of Piezoelectric Characterisation using Resonance and High Field Measurements

M. Stewart and M. Cain

National Physical Laboratory, Teddington, Middlesex TW11 0LW

ABSTRACT

The resonance measurement technique has been compared with high field characterisation to examine the validity of the low field approach in predicting in service behaviour. For a hard PZT, the low field response at any temperature was found to be within 1% of the high field value, and the room temperature result within 5% of the values up to 120°C. For a soft PZT composition, the strong applied field dependence cannot be predicted from low field characterisation, and the material frequency dependence leads to an underestimation of d_{33}.

INTRODUCTION

With the increasing demand for fast and precise movements in many industries, the use of piezoelectric materials, in particular PZT ceramics, has widened. In order to get more movement and decreased response times the materials are being pushed harder, further away from traditional in service conditions. It is well documented that these materials are non-linear with respect to applied field and stress. For most sensor applications the stress levels are such that non-linearity is not a primary concern, however for actuator applications it can lead to large decreases in efficiency due to increased loss levels, or give massive hysteresis in positioning. Many applications of piezoelectric actuators are confined to scientific apparatus where temperature ranges are limited. However, increasing usage in harsh industrial environments has resulted in a need for materials characterisation covering a wider temperature range.

Making measurements at high stress levels can be difficult and time consuming, whereas simple electrical measurements, such as 1 kHz capacitance and loss or resonance, are routine. Consequently, manufacturers will often test components using these simpler methods and it is not until final assembly that the material performance at the desired stress levels is examined. This work examines the validity of this approach, i.e. does the low field behaviour accurately reflect the behaviour seen in real applications?

Characterisation methods for piezoelectric materials are often divided into two categories, low field and high field, based on the applied signal level of the measuring instrumentation. Low field or small signal characterisation is usually associated with impedance analysers that supply a maximum 1 V rms output, whereas high field measurements use amplifiers capable of voltages orders of magnitude greater. The distinction is in fact somewhat arbitrary since it is possible to get non-linear behaviour with 1 V applied on a 0.5 mm thick sample, and the distinction should be the presence of non-linear behaviour in the experiment.

EXPERIMENTAL

Samples suitable for performing length mode resonance were supplied in a range of PZT materials by Morgans Electroceramics. PC5H and PC4D cylinders 15 mm long, 3 mm diameter, were used for the resonance and later for the displacement measurements.

Impedance sweeps were carried out over the relevant frequency range for each sample using an HP4294A impedance analyser. Compensation for cable and fixtures was performed by open and short compensation over the required frequency range. The samples were held at the centre of each end by small spring loaded probes. In order to do the resonance measurements over a range of temperatures the sample holder was placed on a fixture that it could pass through the side entry port of the environmental chamber. A minor modification to the system was the use of coaxial cable rated to 200°C, after lower rated cable shorted due to softening of the sleeving.

The environmental chamber (Design Environmental BS120) has a temperature range of –65°C to +130°C, using liquid nitrogen cooling. Labview software was written to perform the impedance sweeps and also to control the chamber temperature. Since measurement of the sample with a thermocouple would cause significant dampening of the resonance, the chamber thermocouple was used to measure temperature. However, the measurements were performed after five minutes stabilisation at the desired temperature. This practice was verified by independent tests on non-resonating samples with thermocouples touching their surface.

The resonance curves were processed using piezoelectric resonance analysis software (PRAP). In addition to performing the standard IEEE calculations to determine the relevant materials' parameters it can also derive these parameters in complex format which can give some idea of the losses. PRAP uses all of the resonance curve data, not just the maxima and minima, to produce a theoretical fit based on the treatment of the materials parameters as complex quantities. Obviously the width of the peaks contains important information on material losses, which is discarded in the standard IEEE analysis

The length mode resonator enables the determination of several materials parameters, including the piezoelectric coefficient d_{33}. d_{33} can also be obtained by measuring the strain/ displacement of the same sample under an applied field. A system to make these type of measurements was set up using a fibre optic displacement sensor, but after obtaining inconsistent results when performing the measurements at temperature, a capacitance displacement sensor was used instead. A schematic of the system is shown in Fig. 1. The complete

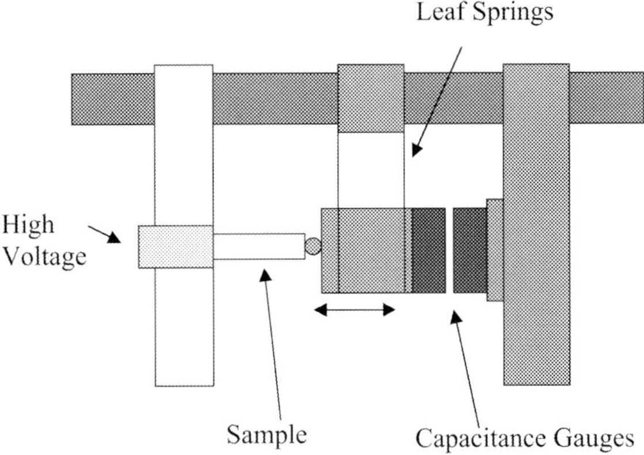

Fig. 1   Schematic of indirect d_{33} measurement system using a capacitance displacement sensor.

system was again fabricated to enable insertion into the environmental chamber. In order to improve the signal/noise ratio a Solartron 1260 Gain Phase analyser was used to measure d_{33}. The output of the Solartron was amplified 150 times using a TREK 50/750 amplifier, and the analyser then measures the d_{33}, dividing the displacement sensor output (Queensgate NS2000) by the applied voltage. Again, the thermocouple in the environmental chamber was used to define the measurement temperature. The samples were left at the desired temperature for 5 min before a measurement was performed. Interestingly, at high field, although the temperature stabilisation was achieved in 5 min, the first set of measurements at each temperature were very noisy. However, when the sample/system had been 'exercised' by applying several high voltage cycles, the measurements at that temperature stabilised. A subsequent increase in the temperature gave rise to more noise, until the sample had seen several high voltage cycles. It may be that this measurement is system rather than material related although similar behaviour occurs when applying static loads to PZT materials under cyclic fields, i.e. there is a change in length after applying the first few field cycles, as the domains seek more stable orientations.

RESULTS AND DISCUSSION

Some of the results from the resonance experiments as a function of temperature are shown in Figs 2–4. The d_{33} for the soft composition shows a linear increase over the measurement temperature range, and the clamped permittivity, ε_{33}, also shows a similarly large increase, if not quite as linear. Also, the results for the heating and cooling cycles are almost identical, confirming that there is no degradation of the sample during the thermal cycling, and that the 5 min sample temperature stabilisation time is sufficient. In contrast, the hard composition shows much less temperature sensitivity (Fig. 3), and there is also some hysteresis as the d_{33}

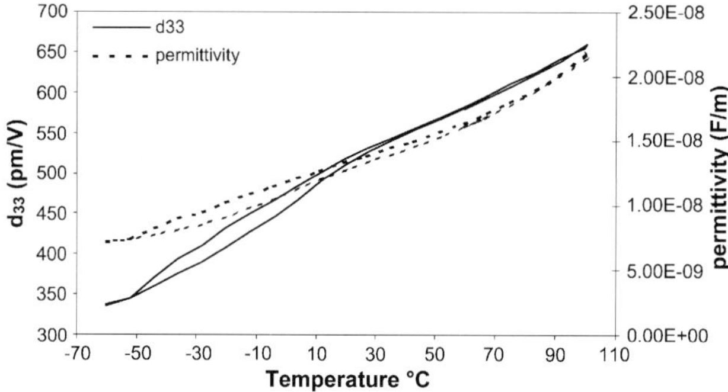

Fig. 2   Piezoelecric characterisation of soft PC5H material as a function of temperature by resonance method.

value on return to room temperature is slightly higher than before. This is due to de-aging of the hard composition during thermal cycling (all the samples used were aged). A second thermal cycle shows more repeatable behaviour. As discussed, the PRAP analysis software can also determine all the materials parameters in a complex form, for example Fig. 4 shows the mechanical loss of the hard and soft materials. The loss is broadly constant over the temperature range, however there is a large anomaly upon going below 60°C on the downward temperature sweep. Upon investigation this feature was found to be present in the compliance results on all the samples, and was eventually traced to the sample holder. One of the spring holders was sticking on the cooling cycle, giving rise to an increased mechanical loss, and when the component was replaced this anomaly disappeared. The values for the lossy parts of the coefficients determined by the PRAP method are consistent with expected behaviour, i.e. the soft materials are generally an order of magnitude greater than the hard material, and the values also agree with the determination of mechanical Q using the peak width. However it is experimentally difficult to measure independently these losses by other means.

The resonance measurements were all performed at 500 mV, giving an applied field of 0.03 V mm^{-1}, and in the analysis it is assumed that the measurements are independent of

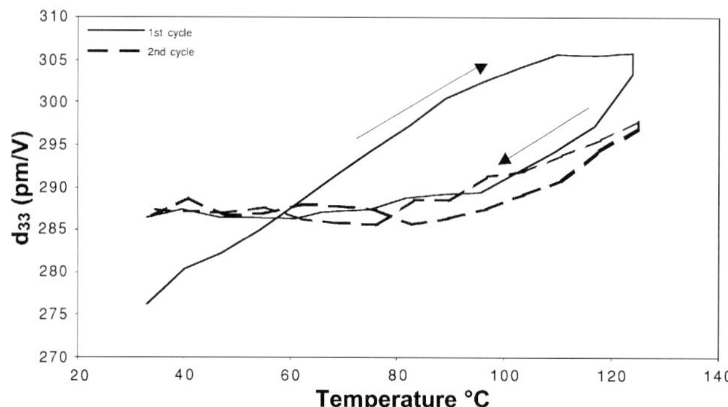

Fig. 3   Piezoelectric coefficient d_{33} for hard PC4D material as a function of temperature.

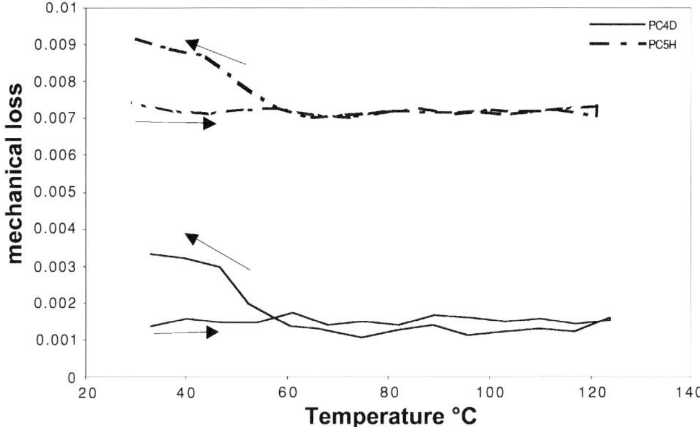

Fig. 4 Elastic loss as a function of temperature for PC5H and PC4D as determined by resonance and PRAP software.

applied field at this level. In contrast, for the displacement measurements at fields from 1.5 to 30 V mm^{-1} there is a large field dependence (Fig. 5). The soft material shows a highly linear increase in d_{33} with increasing field, and increasing temperature also shows increased piezoelectric response, whereas the response of the hard material is much less sensitive to

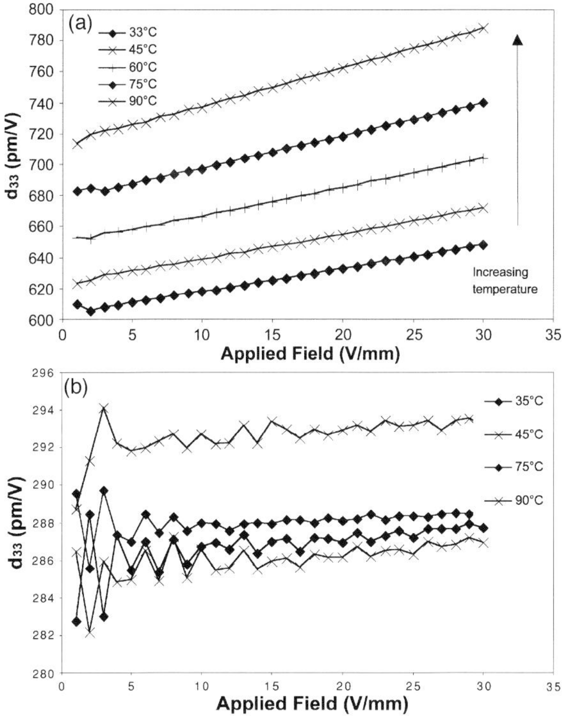

Fig. 5 d_{33} against applied field at different temperature for (a) PC5H and (b) PC4D.

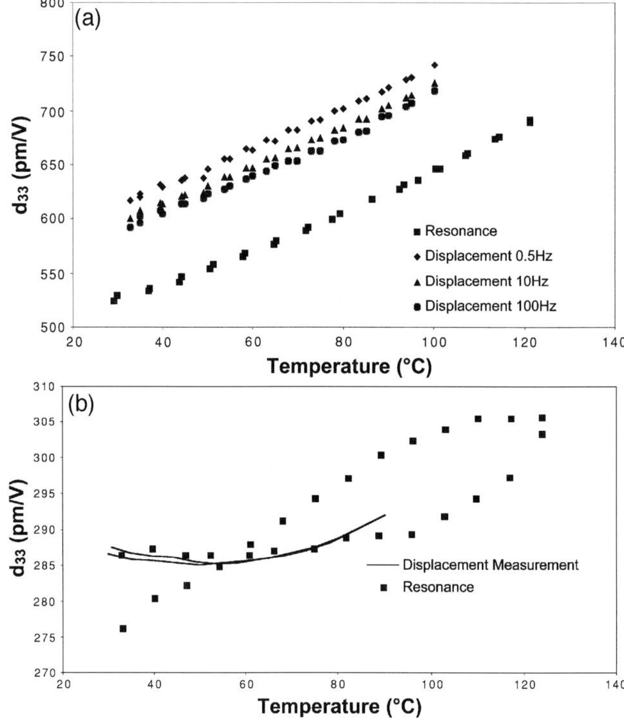

Fig. 6   Comparison of d_{33} measurements by resonance with indirect determination for a range of temperautres. (a) PC5H and (b) PC4D.

temperature and field. The results at low field tend to be noisier because at this level the piezoelectric displacement is only around 1 nm. In order to compare these results to the resonance measurements, the high field results can be interpolated back to low applied fields where the field dependence is negligible. The simplest approach assumes a linear field response and thus the point where the d_{33} versus field curve intercepts the Y axis should be equivalent to the resonance results. Figure 6 compares the high field and resonance measurements for the hard and soft materials. The hard material shows excellent agreement between the two, although the de-aging effect is not seen in the displacement measurements. This is because the high fields applied during the measurement process de-age the samples. The temperature response of the soft material is almost identical for the two measurement methods, apart from the large offset of the high field measurements. Because of the good agreement of the measurements for the hard material, the discrepancy for the soft material is most likely to be material related rather than experimental. There are two possible reasons for the difference, both due to erroneous assumptions in the measurement theory. Either the soft material is strongly frequency dependent, or the linear interpolation of the field versus d_{33} response is invalid. Most work on field dependence of piezoelectric behaviour shows a

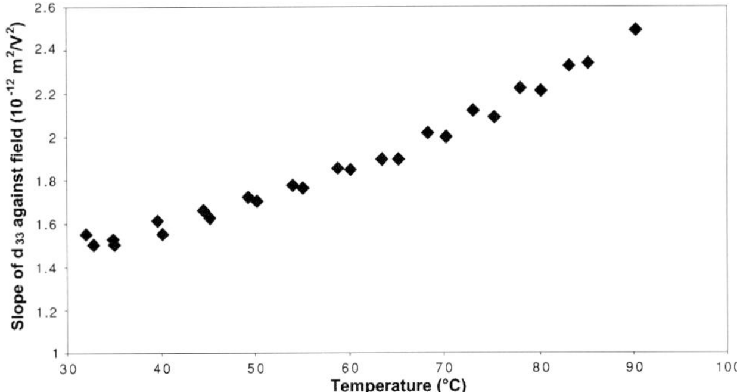

Fig. 7   Change in slope of d_{33} versus applied field responce (10^{-12} m^2 V^{-2}) against temperature.

region of field independence before domain movement contributes to increased response.[1] This would mean that linear interpolation in the field dependant region would give a lower than expected result, whereas we see an increased response. A more likely explanation is the frequency dependence of the soft material, and Fig. 6a shows that increasing frequency decreases the d_{33}. There is much evidence to show that piezoelectric response reduces linearly with the log of the frequency, as the inertia of the domains decreases their contribution at higher frequencies.[2] The resonance measurements were made at several hundred kHz but it is impossible with the current system to increase measurement frequencies much above 1 kHz, because of the response of the capacitance sensor and the mechanical resonance of the measurement assembly.

As can be seen from Fig. 5a, for the soft material, the rate that d_{33} increases with applied field increases with rising temperature, i.e. the slope of d_{33} against field increases with increasing temperature. Figure 7 plots the slope of the d_{33} versus applied field against temperature for the PC5H. The change with temperature is similar to the low field resonance behaviour. The fact that the temperature behaviour is roughly linear gives the opportunity to linearise the curve and thus predict the d_{33} for any given field and temperature. This gives rise to the following empirical equation

$$d_{33} = 1240 + 196380/T + \text{applied field}(\text{V mm}^{-1})*(7.2 - 1749.4/T)$$

which is plotted in Figure 8 along with some experimental data. The first two terms are essentially a linearisation of the low field behaviour, and the third is related to the effect of applied field. Although the equation gives a reasonable agreement with experimental data, the discrepancy arises from the assumption that behaviour is linearly dependant on temperature, whereas, the measured behaviour shows a change in slope around 65°C. The fit could be improved by using a polynomial function, but this refinement is probably not justified considering the errors associated with the measurements. As it stands the equation provides a much improved estimate of piezoelectric response when driving PC5H at reasonable fields and temperatures over the traditional low field characterisation.

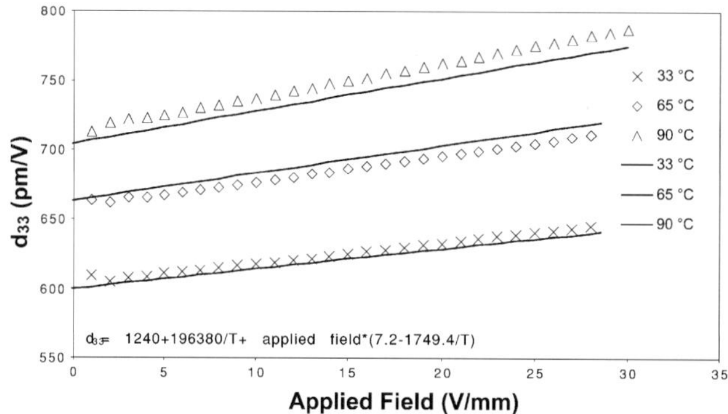

Fig. 8   Experimental d_{33} against applied field for PC5H at several temperatures and fitted response based on the linear equation d_{33}=1240+196380/T+ applied field*(7.2–1749.4/T).

CONCLUSIONS

The response of PC4D with increasing field and temperatures is such that the low field piezoelectric charge coefficient d_{33} at room temperature would give an estimate within 5% for the temperature and field range used in this study. For any given temperature, the low and high field response are within 1% of each other. However, for the soft composition PC5H, although the response with respect to changing temperature is similar at low and high fields, there are two important differences. First, there is a strong field dependence of the soft material that cannot be predicted from low field characterisation. Second, the material exhibits a strong frequency dependence that leads to a large underestimation of d_{33} at low field. Obviously the advantage of the low field characterisation method lies in its simplicity, and the ability to determine many material parameters, which for modelling using FEA is becoming increasingly important.

ACKNOWLEDGEMENTS

This work was carried out under the sponsorship of the UK DTI, part of the Characterisation and Performance of Advanced Materials programme. The samples used in this work were kindly provided by Morgan Electroceramic Ltd.

REFERENCES

1.   D. Hall and P. Stevenson, *Ferroelectrics*, 1999, **228**, 138–158.
2.   D. Damjanovic, *Rep. Prog. Phys.*, 1998, **61**, 1267–1324.